Ohio Hill Country

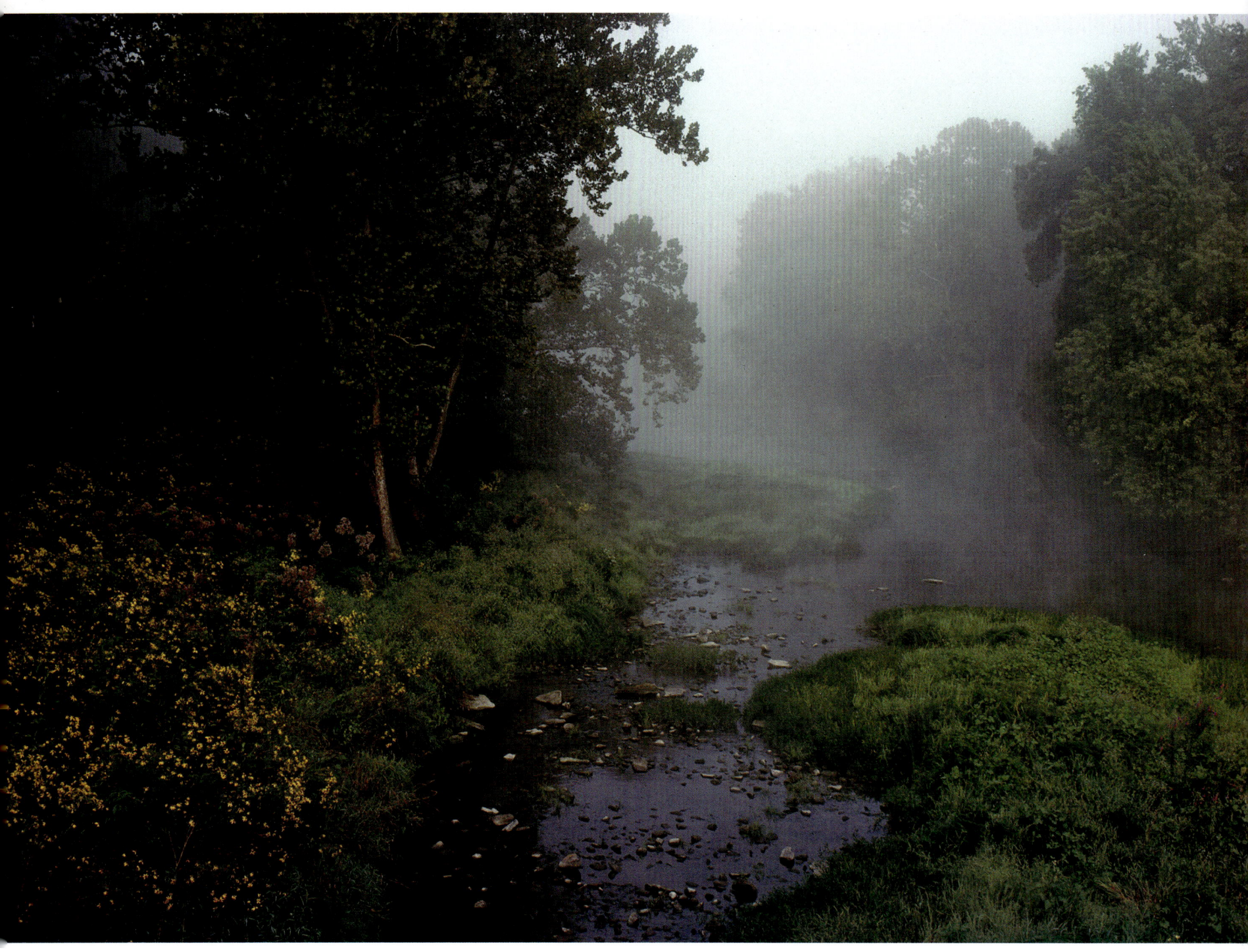

OHIO HILL COUNTRY

A Rewoven Landscape

Carolyn Platt

THE KENT STATE UNIVERSITY PRESS

Kent, Ohio

© 2012 by The Kent State University Press, Kent, Ohio 44242

All rights reserved

ISBN 978-1-60635-134-5

Manufactured in China

Cataloging information for this title is available at the Library of Congress.

16 15 14 13 12   5 4 3 2 1

For my husband, Eric Hoddersen,

a man with a curious mind,

a fighting spirit,

and a very generous heart.

He has been my biggest break in life.

Contents

Preface and Acknowledgments

I grew up in northwestern Ohio, where flat, rich farmland formed in an old bed of Lake Erie. When I was fairly young, my family took a car trip to the Hocking Hills in southeastern Ohio. I remember that vividly, because it was one of the earliest times I saw a countryside so different from my own: the rugged ravines, the massive overhang at Ash Cave, expansive views from the rim at Conkle's Hollow, the thick green pelt of forest, the waterfalls.

As an adult, I wrote quite a few articles for the Ohio Historical Society about Ohio's natural areas, including various places in the state's hilly southeastern third. So when a few years ago, my friend Bob Baris, who lives in the hill country, suggested a book about that part of the state, I was interested. Unfortunately, funding dried up during the recent economic downturn, and the manuscript languished until the Kent State University Press agreed to publish it. I'm pleased and grateful to both Director Will Underwood and the Press's board for helping *The Ohio Hill Country: A Rewoven Landscape* come to light; I also appreciate Dr. Daniel Holm of the Kent State University Department of Geology, who generously read and commented on the manuscript.

Numerous others deserve my thanks as well. As I page through many articles published by *Timeline* editors Chris Duckworth and David Simmons of the Ohio Historical Society, I remember their support and that of their staff. Then there's Gary Meszaros, a fine field biologist, whose great photos and sound advice have made possible not only this book but also my others, *Creatures of Change: An Album of Ohio Animals, Birds of the Lake Erie Region,* and the *Cuyahoga National Park Handbook.* I think of my mother, who infected her family with love of the outdoors, and my father, who was intrigued by geology. Historians both, they sparked my interest in relating human history to that of the natural world.

I think of the naturalists at the Cleveland Museum of Natural History and the Holden Arboretum, who gave me much of my ecological education. There were the museum librarians, especially Wendy Wasman, who made sure I found the information I needed, as well as many staff members at the Ohio Department of Natural Resources and the National Park Service, particularly Jennie Vasarhelyi, who shared information gladly in the midst of busy schedules. There are too many people to name individually, but I remember and thank you all as I begin a new life in the mountains of southern Oregon.

Ohio Hill Country

On a damp April afternoon, the sun comes out to gleam through the passenger window as the car heads southeast from Columbus in central Ohio, through Lancaster, toward Logan on U.S. Highway 33. The land sweeps gently on both sides of the road, neatly squared into rich fields that will rustle with green corn in late July; the Hocking River slides tamely along the wide valley floor to the left. This poorly drained farmland is fruitful, planed smooth by ice-age glaciers and enriched by centuries of elm-ash swamp forest before the early settlers took it up. It speaks of fertility and long-established farm culture.

Suddenly, the country begins to change into its own opposite. Contours roughen as the road crosses that point near Sugar Grove where the vast Illinoian glacier finally ceased its invasion and melted back to the north about 100,000 years ago. Bedrock hills appear on the horizon, marching steadily along the borders of the gently molded glacial outwash valley. They are sharply cut into deep, cool ravines shadowed by giant hemlocks, into massive sandstone cliffs above fast-running creeks, and into dry, sandy ridgetops swept by cold winds in winter and beaten by summer's hot sun. Anyone with an eye for landscape senses that an important transition has been made, and indeed it has, from the low, relatively flat continental interior to the dissected layers of the Allegheny Plateau, "the foothills of the Appalachians," from glaciated Ohio to unglaciated hill country.

Drivers may notice other marked transitions at this Appalachian Escarpment that runs through Ohio from northeast to southwest and continues through Kentucky and Tennessee and into Alabama. Hills rise abruptly, for example, above the valley of Ohio Brush Creek in Adams County east of Cincinnati on the Ohio River. In northeastern Ohio, a long grade on state Route 7 in Columbiana County descends a terminal moraine of the great glaciers into the hill country; here, one has the slightly disorienting sense of rolling *downward* into the hills. From the air, Ohio's rough southeastern third contrasts with the relative evenness of the remainder and claims kinship with Appalachian states like West Virginia and Pennsylvania to the east.

Hocking Hills' cliffs and overhangs are certainly grand, as are other terrains in southeastern Ohio. However, these hills are not the Himalayas, the Alps, or the Rockies, landscapes whose sheer massiveness immediately grasps attention. Ohio hill country asks for a closer look, one that perceives the deep history of this ancient place. It is a history that interleaves eons of extreme mountain building with ages of quiet deposition in shallow seas and in vast prehistoric

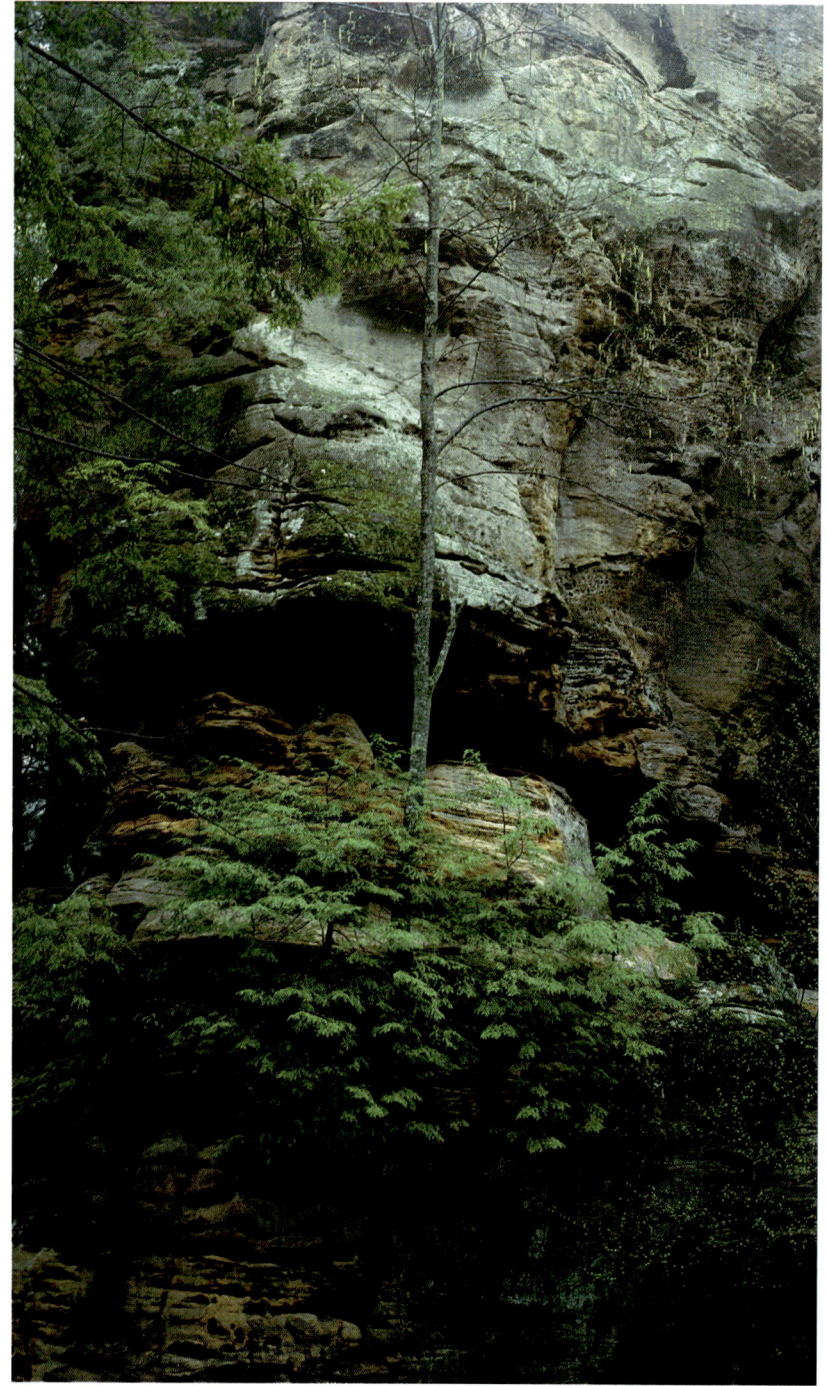

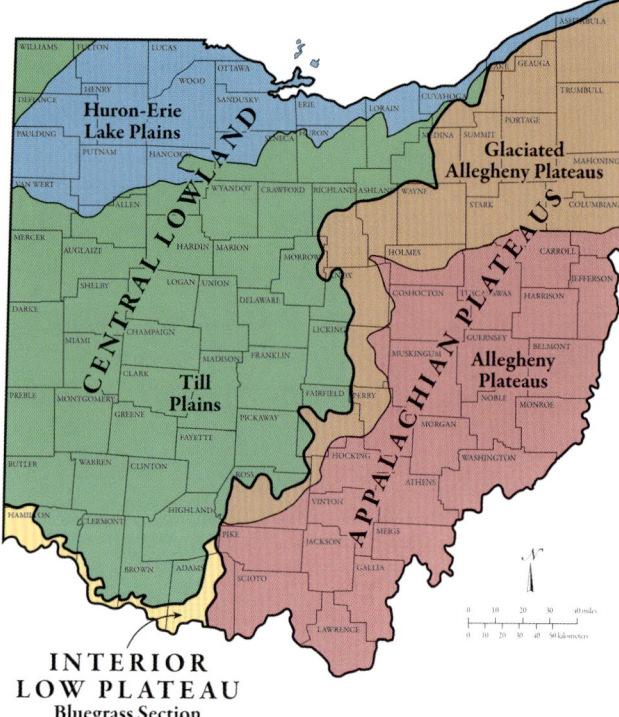

Left: A steep, shadowed valley, dark hemlocks, and diverse deciduous trees create moist green coolness near Cedar Falls in the Hocking Hills. On a hot summer day, hill country gorges offer relief from the heat and provide a home for many ferns, including common polypody, maidenhair fern, and others. *Gary Meszaros*

Above: Five natural regions form modern Ohio: the lake plain, the till plain, the bluegrass region, the glaciated Allegheny plateau, and the unglaciated Allegheny plateau, also known as the Ohio hill country. Ice-age glaciers stalled at the hill country's edges primarily because its bedrock layers were more resistant than those to the north and west. *Wild Ohio*

swamps. It includes extended erosion that erased large chunks of local prehistory and, finally, stupendous continental glaciers that sent overwhelming torrents of meltwater to scour river channels and re-arrange the whole area's river system.

Not only is the variously eroded landscape a product of all these forces, but so is its clothing, the diverse vegetation that covers its contours. Most rock layers in this part of the state are sandstones and shales, which produce acidic soils supporting a very different collection of plants than do limestone and dolostone-derived "sweet" soils to the north and west. Animals, in turn, are adapted to these differing plant communities; for example, the black racer inhabits hill country, while another species of snake, the blue racer, dominates in glaciated parts of the state. The Ohio hills have also been heavily marked by human dramas; by axes of European settlers that obscured the relatively light touch of Native Americans; by canals, mills, and their dams; by timber-hungry iron furnaces of the nineteenth-century Hanging Rock Iron Region; by clay and gravel excavations; by gas and oil wells; and especially by coal mines, both underground and on the surface.

The hill country today is more wooded than the rest of Ohio because its steep ravines and shale and sandstone bedrock have supported poorer farms than the rich fields that now stretch westward and

Landscape and exposed bedrock in the unglaciated southeastern third of Ohio differ from those to the north and west, as do the area's plant and animal communities. This sleek black racer is a denizen of the hill country; its close relative, the blue racer, actually a gunmetal gray with greenish overtones, replaces it in western Ohio. *Gary Meszaros*

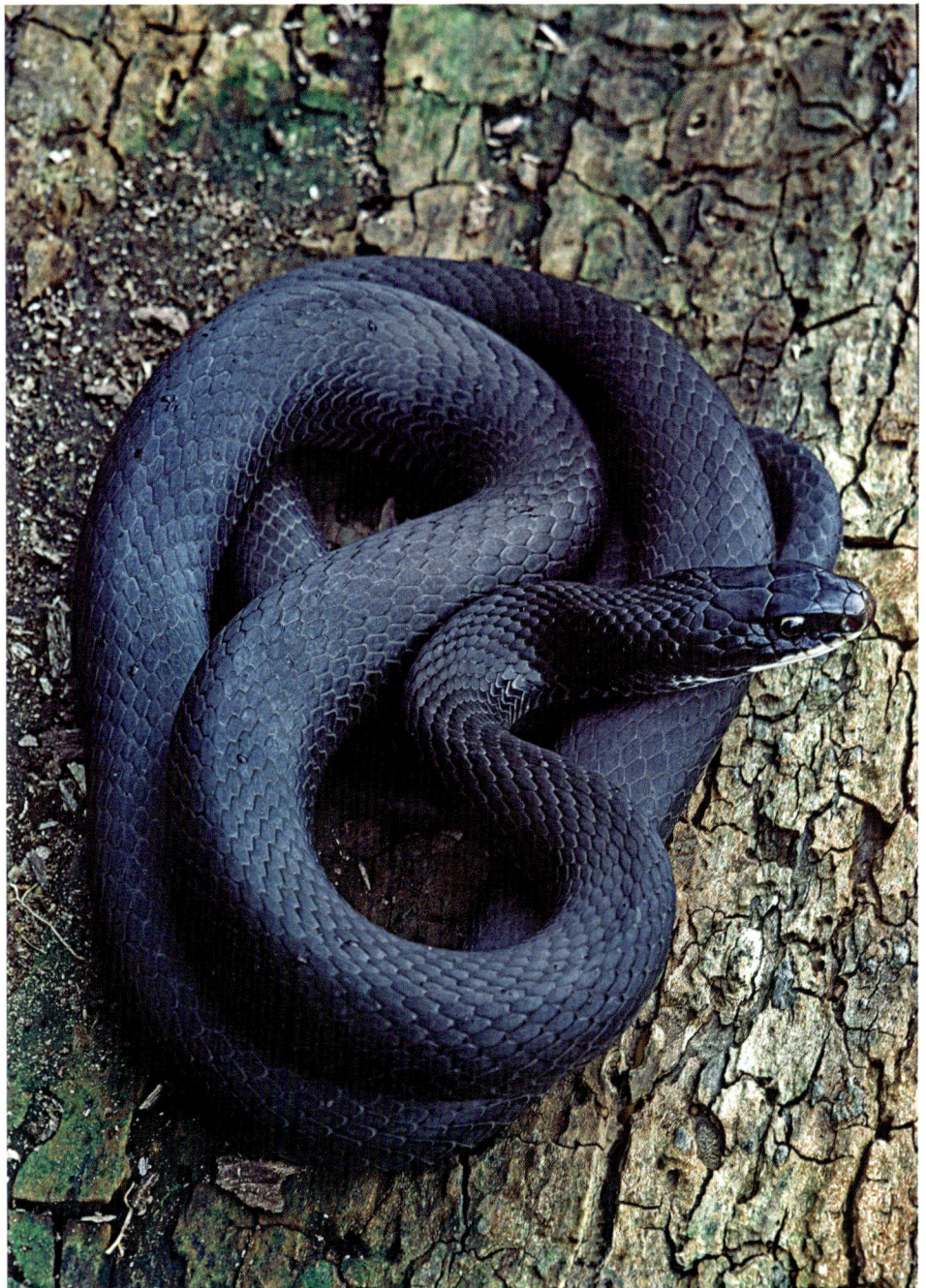

3

northward where the glaciers once lay. Many farmers abandoned their hardscrabble acres in the early twentieth century. In fact, Tar Hollow State Park and Forest near Chillicothe was originally part of a Depression-era program to relocate families on better land where they might survive as farmers. The program ended, and the land finally transferred to the state in 1958.

Because of these displacements, old fields have grown again into woodlands in many areas of the hills, though altered from the original virgin forests. If one drives south from Carroll County, whose rolling pastureland still speaks of its heyday as a dairying center, into western Jefferson County, the transition to wooded hills is quite striking; the little town of Bergholz greets drivers with the sign, "Wooded Hills, Warm Hearts," and the town's name itself comes from the German words for "mountain" and "wood." The Ohio hill country, shaded by recovering forests and deeply dissected by streams, may be poorer agriculturally than many more prosperous parts of the state, but it is beautiful, more varied, and, in many ways, more interesting. It is an old and complex landscape with intriguing geological and ecological histories.

Ancient Ranges

Ohio, and especially this part of the state, is a countryside defined by mountains—not only by the Appalachians but also by three older ranges that rose repeatedly over hundreds of millions of years along the eastern edge of the continent. These were the Grenville, the Taconic, and the Acadian mountains. All four ranges were caused by proto–North America's collisions with other landmasses as tectonic plates ferried the continents and the beds of primeval seas atop Earth's plastically flowing asthenosphere.

Oldest and possibly biggest were the Himalaya-sized Grenville Mountains. They rose when a nameless continent (probably Baltica) collided with this one between 1.2 and 1 billion years ago to form a primordial supercontinent that we have dubbed Rodinia. Sea floor sediments were squeezed between the two behemoths and had nowhere to go but up. These mountains are the only ones whose eroded roots actually underlie our state because the continent's eastern margin probably corresponded with eastern Ohio at that long, long-ago time. This was once truly the edge of the world.

The western boundary of what is called the Grenville Province, named the Grenville Front, roughly parallels route I-75 from Toledo to Cincinnati. Land west of this line lies over a 1.3-billion-year-old rift

system that failed to break up the continent then but today can still visit Ohioans with earthquakes. That area has quite different underlying rocks. No eager geology buff will ever find rock from the Grenville Mountains at the surface; deposits from Precambrian times lie deep in Ohio's "basement," buried by more recent layers that are thousands of feet thick.

Nothing is constant, it is said, but change, and in the infinitely slow dance of the continents, Rodinia pulled apart, forming the Iapetus Ocean, an ancestor of our present Atlantic. Sediments from the eroding Grenvilles formed a wide continental shelf in the ocean during the Cambrian Period, a shelf that seethed with newly diverse marine life. About 542 million years ago, a shallow sea also submerged ancient North America's interior west of the mountains and covered most of what would be Ohio for nearly 250 million years more. This is lucky for us, since seas leave rock strata and fossil records, while erosion of dry land wipes them out. Trilobites reached their peak diversity late in the period. Unfortunately for fossil hunters, however, Cambrian rocks and fossils are also buried far beneath the surface in this state.

Meanwhile, the heavier ocean crust was slowly diving, or subducting, beneath the east coast of Laurentia, as proto–North America is sometimes

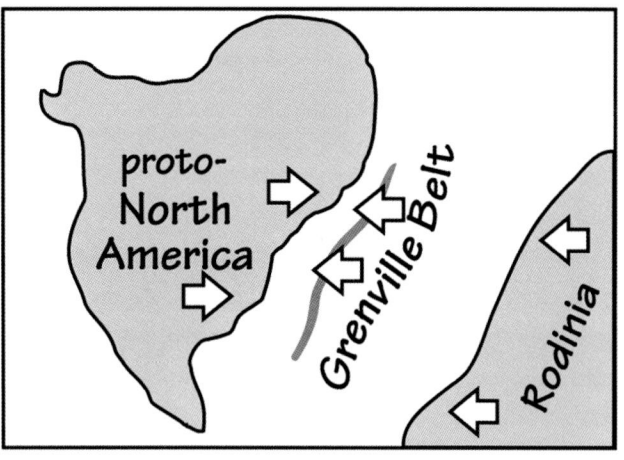

More than a billion years ago, the Precambrian shield, this continent's stable core, collided with one or more other continents to form a supercontinent called Rodinia. Sediments that had earlier eroded from the shield were squeezed between the colliding land masses and subjected to tremendous pressure and heat. They rose to form the massive Grenville mountain range. This was only the first of four such slow-motion continental crashes that helped shape today's landscape. *Paleontological Research Institution*

called, though in those days, the coast actually oriented toward the south. On the crust rode a chain of volcanic islands, the Piedmont Terrane, which plowed slowly into the Laurentian coast during the Ordovician Period about 475 million years ago, along with mixed sediments scraped from the ocean floor. The resulting Taconic Highlands were as forbidding and desolate as the Grenvilles had been, with no tinges of green, since land plants and animals were still millions of years in the future. These mountains eventually stretched from present-day Georgia to Newfoundland. Importantly for our area, the monumental pressures caused by this leisurely crash caused a downward sag west of the highlands that became the Appalachian Basin, vital to the hill

country's development. Sediments from the eroding mountains collected in thick layers in this crustal downwarp. Though Ordovician Period bedrock is mostly buried in Ohio, outcrops do appear at the surface around Cincinnati, an area famous worldwide for spectacular invertebrate fossils.

Why should we concern ourselves with mountain ranges that long ago weathered out of existence? There are very good reasons: Erosion of these mountains largely determined what kinds of sediments were laid down in the inland sea to their west. These layers, in turn, governed the types of sedimentary rocks formed, their resistance to weathering, and the look of landscapes that formed from them, as well as Ohio resources such as oil, gas, salt, coal, and clay for ceramics and bricks. High mountains spawn rushing rivers, and these

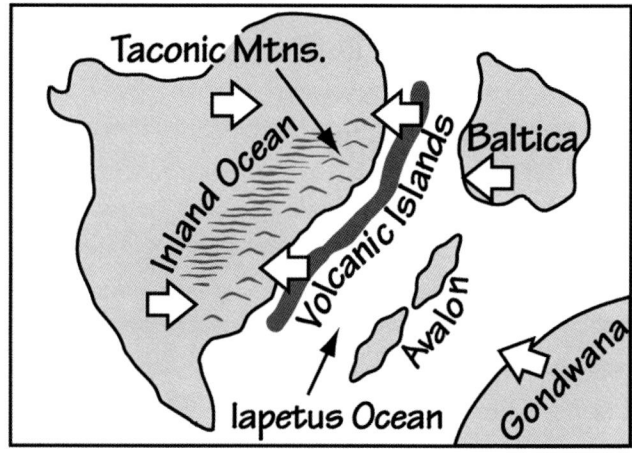

During the Ordovician Period, about 475 million years ago, the Piedmont Terrane, a chain of volcanic islands, docked against proto–North America along with seafloor sediments, to raise the Taconic Highlands. On their western fringe developed a great depression called the Appalachian Basin where thick sediments collected for hundreds of millions of years. *Paleontological Research Institution*

Years before present, in millions of years	Eras and duration in years	Periods and and duration in years	Area of outcrop in Ohio and principal rock types	
	CENOZOIC 66+ million	QUATERNARY 1.5-2 million	northwestern $^2/_3$ of Ohio— unconsolidated sand, gravel, clay	
1.6		TERTIARY 62.5 million		
66.4	MESOZOIC 179 million	CRETACEOUS 78 million	NOT PRESENT IN OHIO	
144		JURASSIC 64 million		
208		TRIASSIC 37 million		
245	PALEOZOIC 325 million	PERMIAN 41 million	southeastern Ohio—shale, sandstone, coal, clay, limestone	
286		PENNSYLVANIAN 34 million	eastern Ohio—shale, sandstone, coal, clay, limestone	
320		MISSISSIPPIAN 40 million	east-central, northeastern, and northwestern-most Ohio—shale, sandstone, limestone	
360		DEVONIAN 48 million	central, northeastern, and northwestern Ohio—shale, limestone	
408		SILURIAN 30 million	western Ohio—dolomite, limestone, shale	
438		ORDOVICIAN 67 million	southwestern Ohio—shale, limestone	
505		CAMBRIAN 65 million	NOT EXPOSED IN OHIO	Cambrian sandstones, shales, and carbonates and
570		PRECAMBRIAN 3,400 million		Precambrian sedimentary, igneous, and metamorphic rocks present in subsurface

Divisions of geologic time and their representation in Ohio. Wavy lines indicate a major unconformity (period of erosion or nondeposition) in Ohio.

Mountain ranges affected Ohio's geology during several time periods: Precambrian (Grenville Mountains), Ordovician (Taconics), Devonian (Acadians), and finally the Pennsylvanian and Permian (Appalachians). Oldest bedrock visible at the surface is Ordovician-age around Cincinnati, and youngest is late Pennsylvanian and early Permian near the Ohio River in the southeastern part of the state. Erosion has obliterated any bedrock younger than this and has created a long gap in the record. Most rocks visible in the hill country were formed during the Mississippian and Pennsylvanian subperiods. *Ohio Department of Natural Resources*

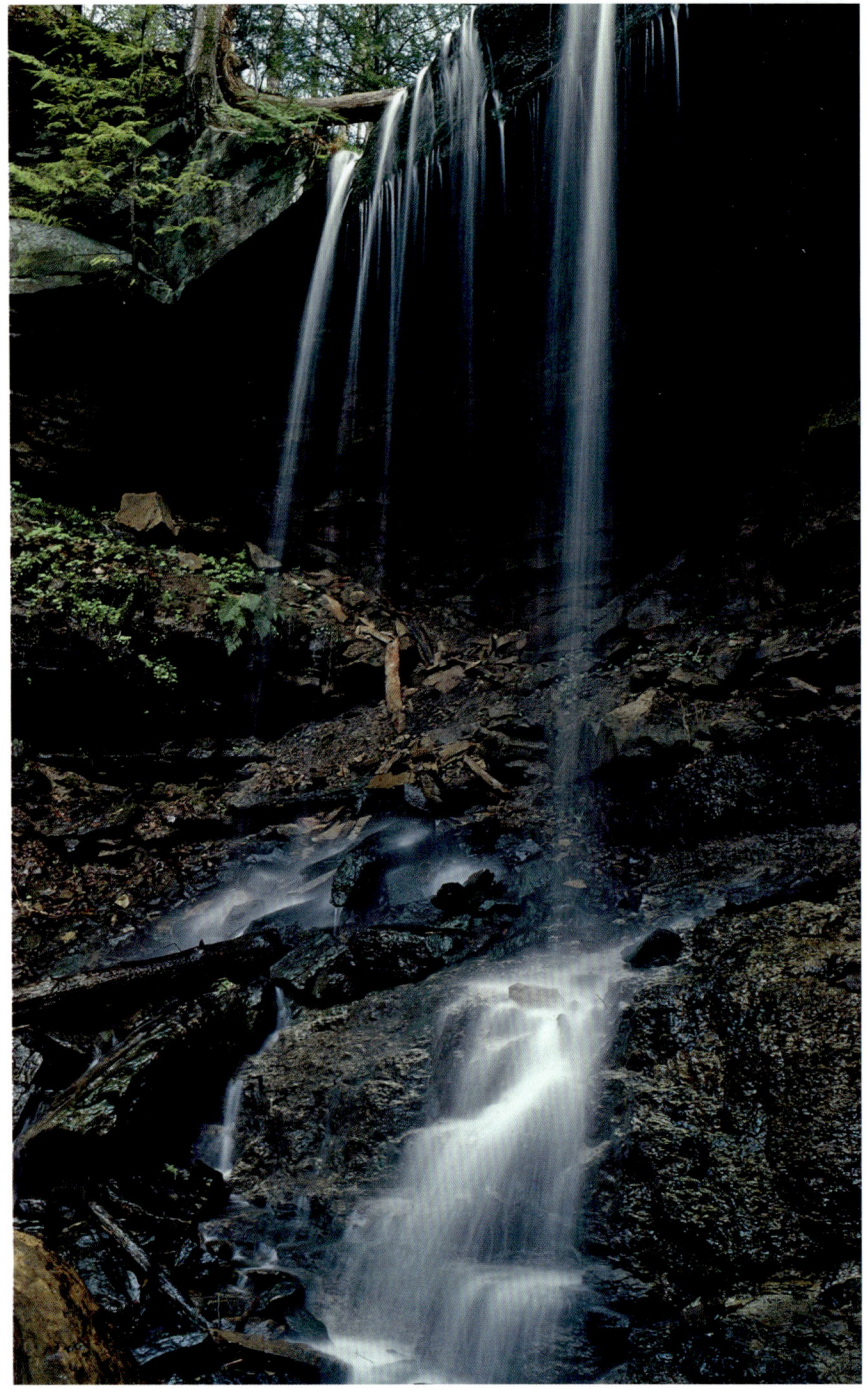

torrents can carry larger particles of rock for longer distances than quiet streams can, so when massifs towered on the seaboard, eastern Ohio received masses of pebbles and sand eroded from them in deltas and braided channels. These eventually formed coarse conglomerate rocks and sandstones. Laid down farther from the mountains were fine-grained clays that transformed into shale under pressure. Still farther west, especially in times without high mountain ranges, warm, clear waters in shallow seas fostered calcium precipitation and collection of shelly materials from ocean life. These became limestones, often later transformed into dolostones as magnesium-rich groundwater percolated through the layers.

This is precisely what happened during the following period of relative quiet, the Silurian. As the Taconics eroded down to rolling but still-barren hills, streams became quieter and carried less silt into the sea to their west. Ohio had lain fairly near the equator at least since Precambrian times, and it continued to do so for hundreds of millions of years. In the warm, limpid Silurian seas, colonial corals formed massive limestone reefs along the Cincinnati and Findlay arches to the west of today's hill country. These and other great reefs blocked circulation of ocean currents, and in northern and eastern Ohio, evaporation created great thicknesses of rock salt and other evaporites such as gypsum. These 410-million-year-old salt beds lie beneath the

Rivers eroding ancient mountain ranges along the coast deposited the shales and sandstones of today's hill country. Now, streams like this one in Columbiana County are eroding these same layers of rock and depositing their materials elsewhere. *Gary Meszaros*

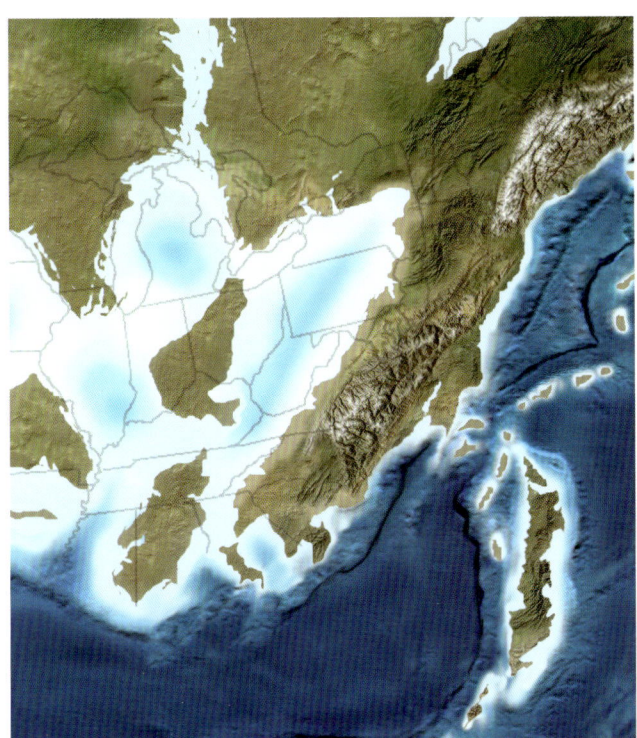

During the middle Devonian, proto-Europe, or Baltica, slowly crashed into this continent; later during the same period, a micro-continent Avalon docked farther south on the coast. The result was a third great mountain range, the Acadians. This re-creation shows the Acadian Mountains and the Appalachian Basin that lay along their western edge. The great downwarp had formed roughly 100 million years before and would persist for another 100 million, accumulating great thicknesses of sediment. Farther west, the map also shows the Cincinnati Arch and the Michigan and Illinois basins. *Colorado Plateau Geosystems, Inc.*

Geologists estimate that reserves amount to an astounding 2.5 trillion tons, which, even if only one-fourth is recoverable, could supply the nation's myriad industrial and culinary needs for the next thirty-two thousand years.

The early Devonian Age of Fishes imitated the halcyon Silurian, though a break in the geological record shows that the inland sea had continued to withdraw from proto-Ohio, erosion making a significant gap in the rock layers. Nemesis, however, was returning for a second time in the form of the Baltica tectonic plate that carried on it what is now called Europe. In another ponderous smashup about 375 million years ago, Baltica collided with the northeastern part of this continent to form yet another range, the Acadian Mountains. Caught between the two behemoths was romantically named Avalon, a micro-continent thought to have strayed from proto-Africa. Later in the Devonian Period, another wandering fragment of Avalon docked along the southeastern seaboard.

Ohio again received sediment from rushing mountain streams that carried future sandstones and shales into the resurgent inland sea, revived by the most recent tectonic commotion. The Berea Sandstone, source of high-quality grindstones for millers of the nineteenth century and many monumental buildings from Chicago to Toronto and Boston, is one result. During the next subperiod, the Mississippian, the Black Hand sandstone, another erosion-resistant layer later used in building, was laid down. This would eventually form the spectacular cliffs, overhangs, and rock shelters of southeastern Ohio's Hocking Hills region. For reasons explained below, Mississippian strata, as well as layers from the next subperiod, the Pennsylvanian,

hill country's northeastern third and extend into northern Ohio, Pennsylvania, West Virginia, New York, Ontario, and Michigan. Pomeroy, in Meigs County on the Ohio River, produced 75 percent of Ohio's salt in the nineteenth century, this when the state was the third saltiest producer in the nation, as it still is today. Now production is centered in Cleveland, largely of salt for highway ice control.

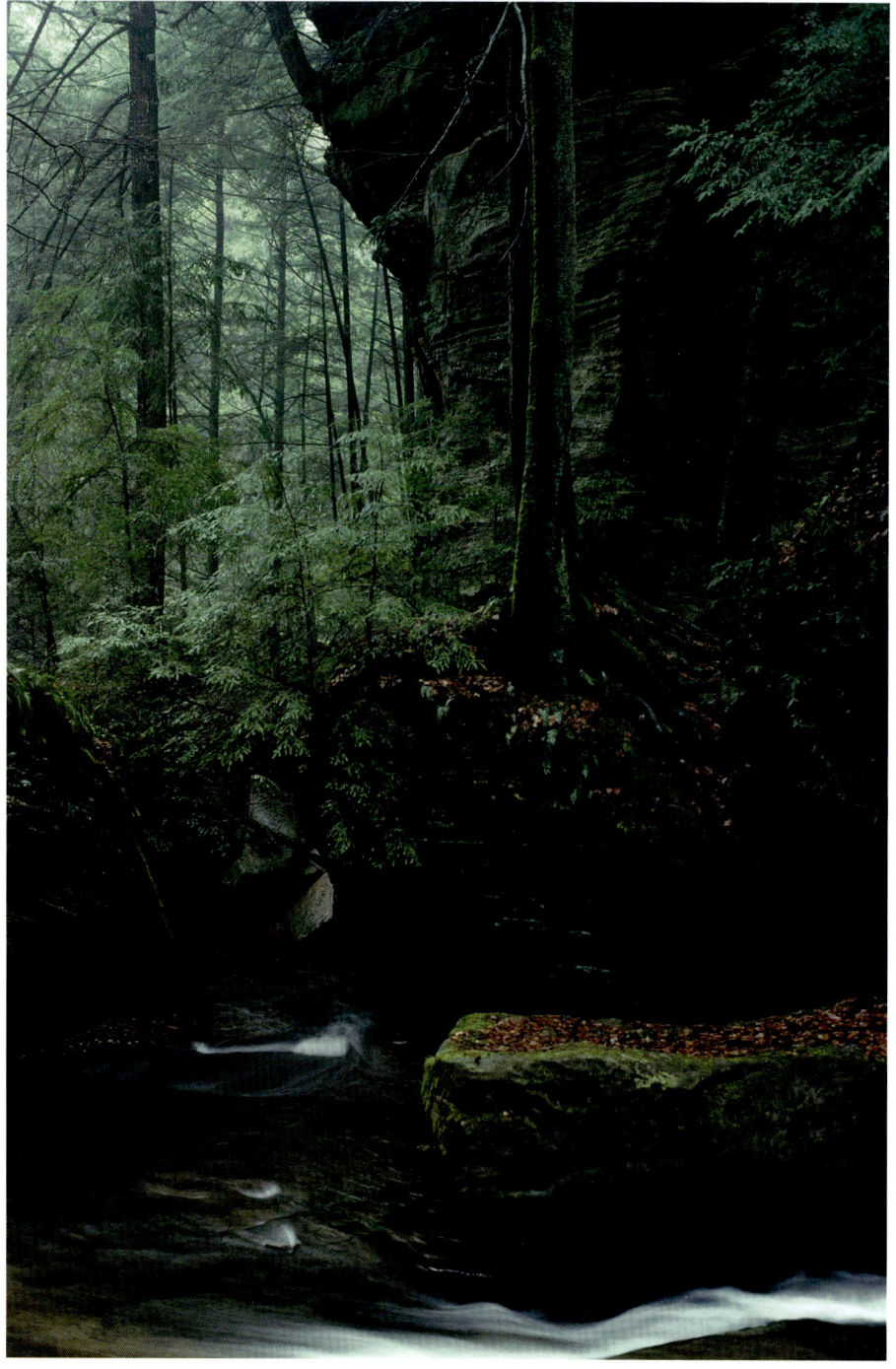

make up most hill country bedrock visible at the surface, along with slightly younger rocks bordering the Ohio River, which may be either late Pennsylvanian or early Permian Period. These hill country exposures vary from roughly 360 to 300 million years old.

In a scene typical of the Hocking Hills' canyonlike river valleys, massive Black Hand sandstone forms a backdrop for lush fern gardens and other undergrowth. Hemlock and yellow birch, both northern trees, grow in the steepest ravines and on shaded slopes. *Gary Meszaros*

Gondwana and the Appalachians

More geological turmoil was to come as Gondwana lumbered at a snail's pace toward united North America–Baltica to bring about another ponderous crash. Gondwana was a giant continent made up of what are now Africa, South America, Australia, India, Antarctica, and, last but not least, Florida. During the Permian Period, about 300 to 250 million years ago, the continents met to form a worthy successor to Rodinia, a gargantuan supercontinent geologists call Pangea, or "all earth."

It was this collision that formed the Appalachian Mountains, which are still with us today, though not at the towering heights they once boasted. The northern Appalachians had roots in the Grenvilles, the Taconics, and the Acadians, all of which had crumpled, folded, and extended the eastern seaboard far beyond the old Precambrian coastline between Youngstown and Marietta. Now pressures from Africa extended this giant range to the south, and today it stretches all the way from Canada to Alabama. A squeeze from proto–South America also created the Oachita Mountains of Arkansas and Texas. Again, fast torrents from the eastern mountains dropped coarser sediments in great deltas with twisting channels at the edges of the inland sea. Pebbly, cross-bedded Sharon conglomerate (sometimes called Sharon sandstone), for example,

forms massive ledges at Hinckley and elsewhere in northeastern Ohio, as well as erosion-resistant knobs in Jackson County near the Scioto River close to the southern tip of the hill country.

Sea levels fluctuated a great deal during Mississippian, Pennsylvanian, and Permian periods,

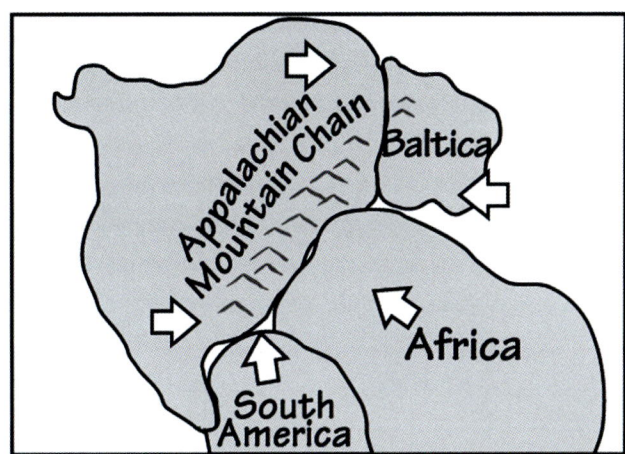

About 300 to 250 million years ago in the Pennsylvanian and Permian periods, Gondwana—a huge continent that included present-day South America, Africa, Australia, Antarctica, and India—collided with North America to form Pangea, a tremendous supercontinent. This caused the most recent mountain-building episode on the ever-accreting eastern continental margin. The resulting Appalachians are still with us today. The great inland ocean that had lain west of the Taconics and Acadians for many millions of years began to recede as the Appalachian Basin filled with sediments from the eroding mountains to the east. *Paleontological Research Institution*

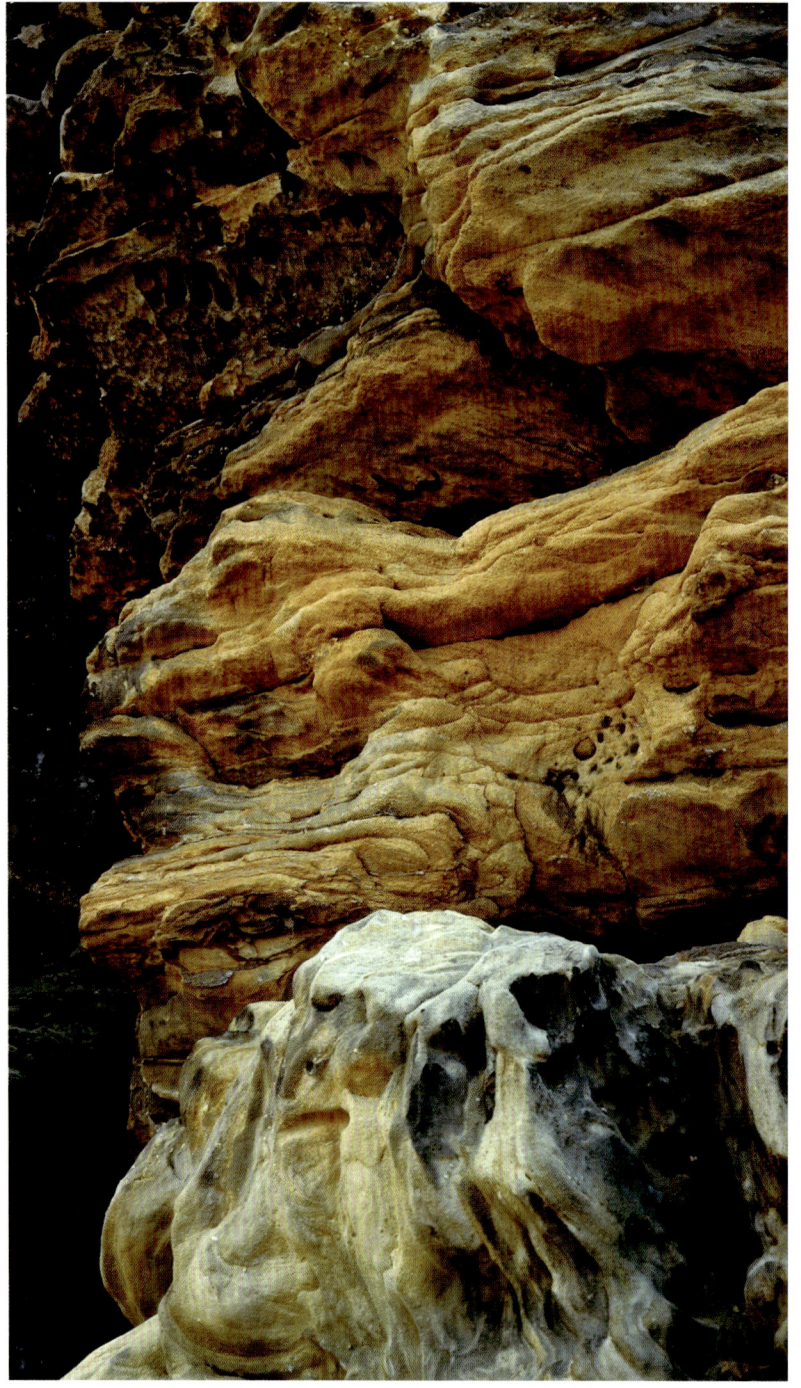

leaving behind varied rock layers. This is why geologists find sea sediments with marine fossils in the hills, deposited when shallow seas made return visits to the area, as well as remains of terrestrial animals and plants from great swamp forests. These in due course compressed to form Ohio's coal measures. Freshwater limestones formed in backwaters and oxbow lakes in the deltas and swamps; underclays often lay beneath coal deposits. The result was a complex assortment of layers with differing thicknesses. Land plants, which began to conquer the continents in the later Silurian, were abundant by Pennsylvanian times, and some, such as tree ferns and club mosses, were quite large. Some giants, whose petrified stumps miners discovered in Jefferson County's Sterling North mine, could have reached 120 to 150 feet tall. Insects, great amphibians, early reptiles, and freshwater clams populated the vast swamps.

Colossal Pangea lasted for 100 million years, but in the end the center could not hold. Plate movements changed over time, and the continents began to rift and pull apart once again, rather, as snacking geologists have observed, like a Snickers bar that slowly cracks and separates under tension. Such rifting is actually happening today in the great East African Rift Valley. Pangea's prolonged divorce process took its course during the Triassic and Jurassic, the first two periods of the Mesozoic Age of Dinosaurs. The breakup completed about 150 mil-

Fast streams rushing off the early Appalachians formed the Sharon Sandstone. Its layers formed in great deltas and show braided channels and polished granite pebbles carried by powerful currents. Because of the pebbles, this rock is sometimes called the Sharon Conglomerate. *Gary Meszaros*

lion years ago as the continents went their separate ways toward the positions they hold today. Baltica moved east, taking part of the northern Appalachians with it. (They are now the Scottish Highlands.) The Atlantic Ocean was born.

North America began to move away from its ancient flirtation with the equator, leaving snowbirds of the future no choice but to migrate to Florida, which had declined to drift away with its old partner Africa during the rifting. India made a long journey north and finally collided with Asia to raise the Himalayas. North America's eastern seaboard became what is termed a *passive continental margin,* and long quiet ages of erosion and deposition created the Atlantic and Gulf coastal plains. Dramatic mountain-building action shifted to the other coast, which had begun to encounter the Pacific Plate to the west as our continent pulled away from Pangea. In Ohio, the geologic record is missing from the early Permian Period until the great continental glaciers began to rumble out of the north a mere 3 million years ago. Ohio has no bedrock deposits younger than about 290 million years of age; the sea had withdrawn, and erosion rather than deposition ruled. Alas, there are no dinosaur fossils in the Ohio hills or in any other part of the state.

The oldest rocks visible at the surface in Ohio are Ordovician (dark pink) and Silurian (light greens) areas on the map. These correspond to the Cincinnati Arch, an ancient Precambrian rift valley that uplifted during the Cambrian Period. Younger rock layers eroded off the exposed back of the arch but can be seen as progressive parallel stripes to the east and in the extreme northeast corner of the state. The most important hill country exposures are from the Mississippian and Pennsylvanian sub-periods. *Ohio Department of Natural Resources*

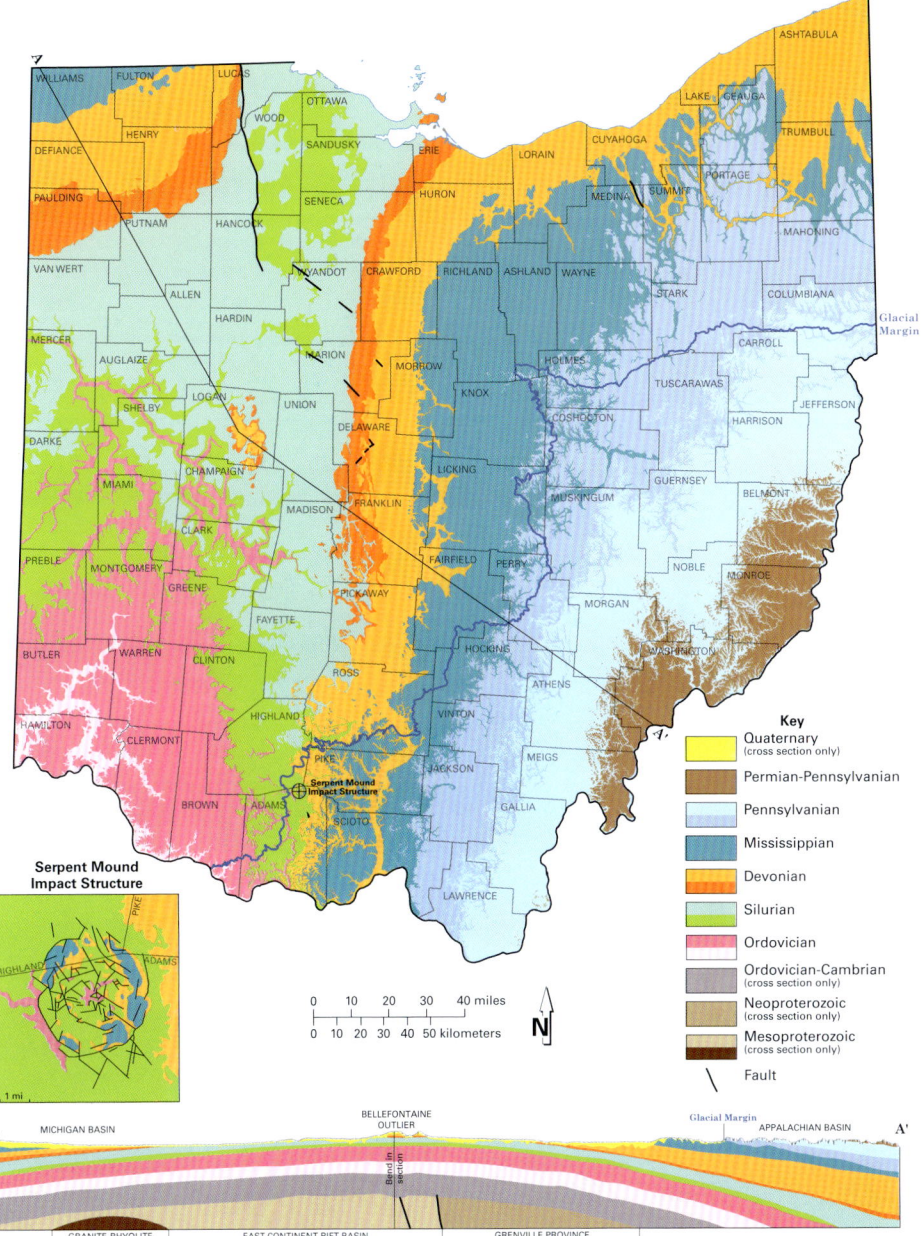

Key

Quaternary (cross section only)

Permian-Pennsylvanian

Pennsylvanian

Mississippian

Devonian

Silurian

Ordovician

Ordovician-Cambrian (cross section only)

Neoproterozoic (cross section only)

Mesoproterozoic (cross section only)

Fault

Erosion, Deposition, and the Ice Age

These clashes of continents explain much of why hill country terrain looks so different from the rest of Ohio. Erosion after the seas withdrew and invasions by the continental glaciers complete the story of today's landscape. If you look at the bedrock map of Ohio, formations from successive periods appear as a striking, northeast-southwest series of bands, generally trending from older outcrops in the western part of Ohio to younger ones in the east. Why should this be so? It seems that Taconic mountain-building in the Ordovician Period caused some parts of the crust, such as the Appalachian Basin, to subside, while others, including the Cincinnati and Findlay arches in southwestern and northwestern Ohio, did not. Sometimes, over millions of years, the arches' backs rose above retreating ancient seas like stupendous sea monsters. During these times, erosion stripped younger strata off their crests, leaving progressively recent layers on their flanks. On the arches, or platforms, between Cincinnati and Toledo, most exposures are of Silurian age; they are bordered progressively to the east

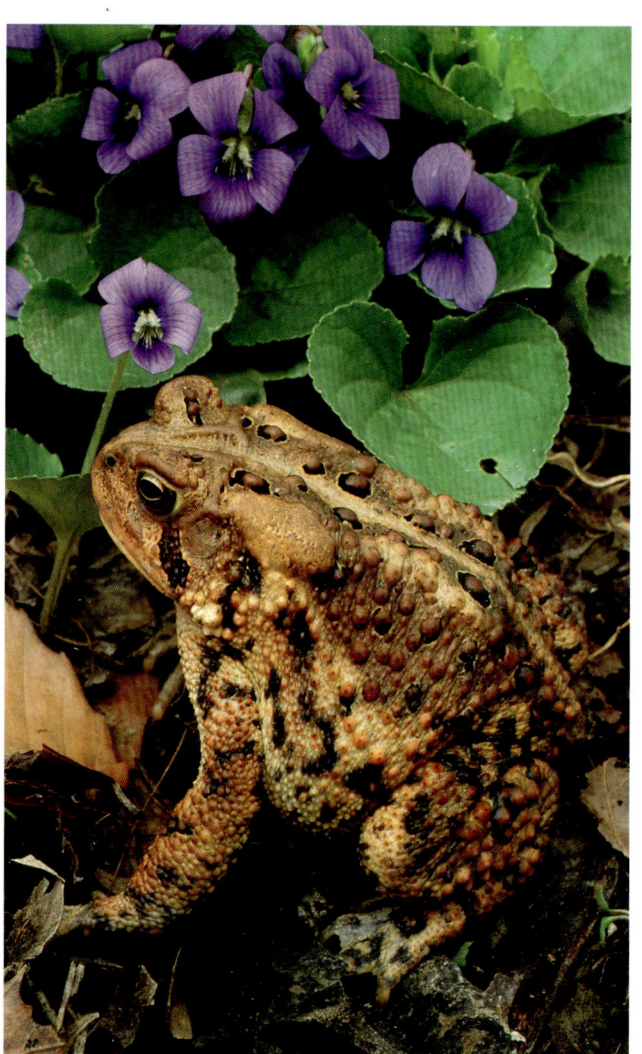

Like many other amphibians, American toads like this one appreciate southeastern Ohio's moist woodlands and vernal pools, necessary for their reproduction. Ohio's geologic history has formed the landscape that sustains them and many other living creatures. *Gary Meszaros*

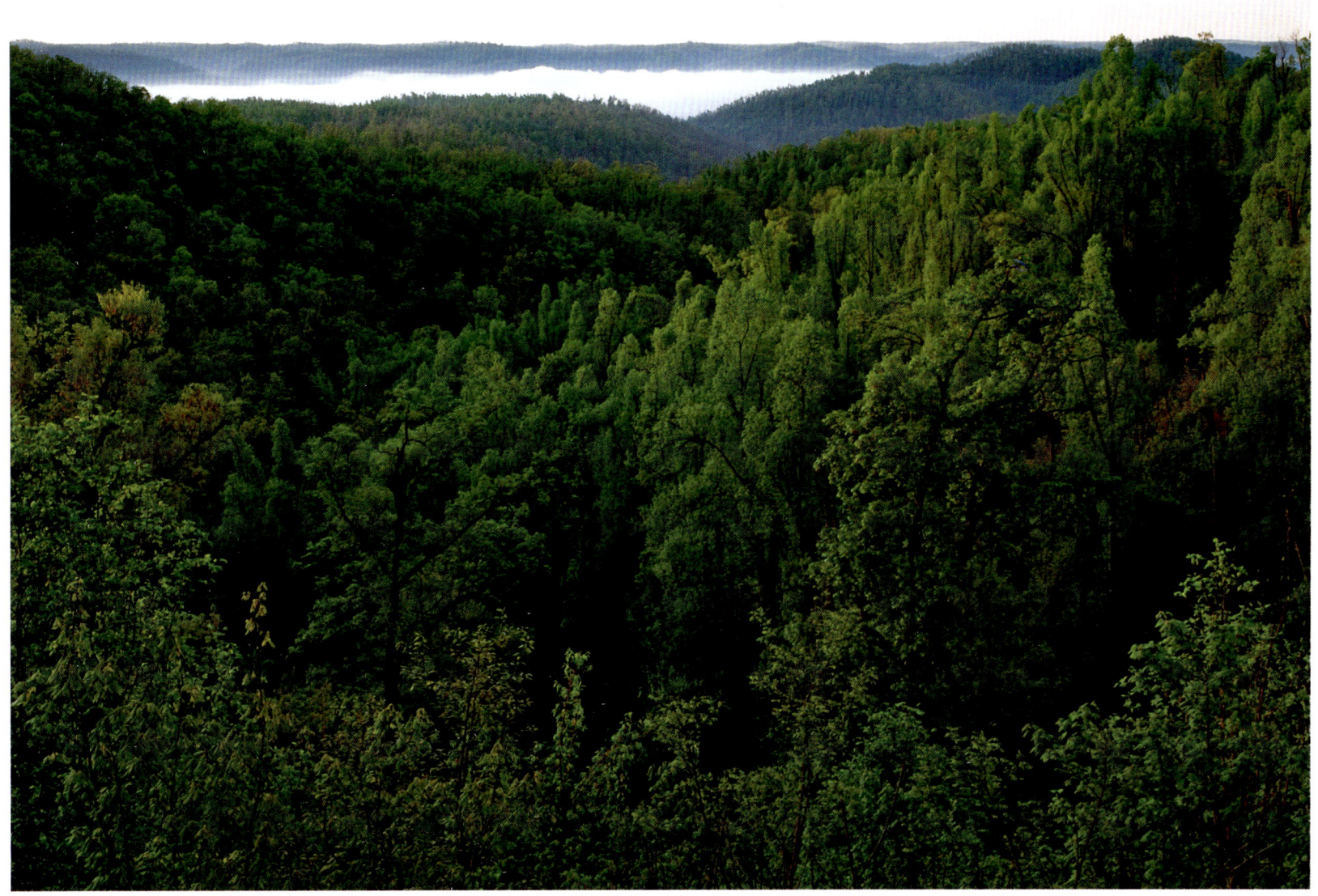

The rolling hills and deeply cut valleys of southeastern Ohio are not the result of uplift but of erosion. Streams have dissected thick layers laid down long ago in the Appalachian Basin, a downwarp in the earth's surface that received sediments from high mountain ranges to the east. At the Ice Age's end, rivers brimmed with prodigious glacial runoff, which deepened valleys and rearranged drainage. This view is of Shawnee State Forest near Ohio's southern tip. *Gary Meszaros*

by strips of Devonian, Mississippian, Pennsylvanian, and early Permian outcrops on the bedrock map.

At the same time, the Appalachian Basin to the east kept on subsiding, which allowed very thick layers of sediments to accumulate; they were so thick that southeastern Ohio is the only place in eastern North American where Permian-age rocks have survived the attacks of wind, water, and frost at the surface. Bedrock maps show that because of these erosional relationships, the hill country is bordered on the west by older Mississippian strata and on the southeast by younger Permian strata, while Pennsylvanian rocks that contain most of Ohio's coal deposits outcrop in a large swathe between the two. Remember that the Acadian Mountains first rose in late Devonian times, about 375 million years ago, and the Appalachians about 318 million years ago, at the boundary between Mississippian and Pennsylvanian times. The Devonian Berea sandstone, the Mississippian Black Hand sandstone and the Pennsylvanian Sharon Conglomerate, as well as other resistant sandstones and shales, are the direct results of deposition from these ranges in the streambeds and deltas of the Appalachian Basin.

When 300 million years ago, the interior seas receded from this area for good, erosion began to fashion thick Appalachian Basin layers into the rolling, deeply dissected landscape seen in the hill country today. Geologists have identified two ancient erosion surfaces to the east and west of an old drainage divide, the Flushing Escarpment, that extends from Columbiana County south through Carroll, Harrison, Belmont, and Monroe counties. The Lexington surface to the Escarpment's west shows relatively low, narrow ridges and fairly broad, shallow valleys.

The Harrisburg surface to the east is higher—the highest hill is 1,388 feet—and intensely cut by deep, narrow valleys. The Flushing Escarpment itself separates streams flowing east to the Ohio River from those flowing west toward the Muskingum.

When the recent Pleistocene, or Ice Age, began, the sandstones and shales of eastern Ohio proved more resistant than softer limestones and dolostones to the west. Because of this, the glaciers stalled at about Canton's latitude in northeastern Ohio. In western parts of the state, they scoured all or nearly all the way to its southern boundaries. As a result of ancient geological history, then, most hill country escaped the glacial planing and scouring that evened land to the west, where the glaciers, as they advanced and retreated, dropped till and moraines that buried most bedrock exposures. European settlers found gentle landscapes and rich farmland in glaciated parts of the Ohio Country.

No such glacial drift hides the rugged outlines of southeastern Ohio's hills, chiseled from the strata laid down eons ago in the Appalachian Basin by runoff from those ancient eroding mountain ranges to the northeast, east, and southeast. Ordinary rainfall, frost weathering, and stream action account for much of this landscape's creation, but not all of it, certainly. Tough sandstones and shales of the hill country may have stopped the big ice sheets where Canton lies today, but the glaciers were hardly so easily dismissed. Although they never succeeded in riding over southeastern Ohio, they rearranged its whole drainage pattern and sent gigantic outpourings of meltwater to shape its landscape further.

Geologists believe that at least four major glacial pulses invaded Ohio, though the last two have destroyed almost all evidence of previous episodes. It seems likely that an early glacier called the Nebraskan annihilated the venerable Teays (rhymes with "ways") River system sometime between 2 million and 690,000 years ago. It did so by damming northward-trending streams with ice to form Lake Tight. Named for Ohio geologist William G. Tight, an early Teays drainage expert, this was New York's Finger Lakes writ very large. Bigger than Lake Erie, whose area is 5,002 square miles, Lake Tight's valley fingers are thought to have covered nearly 7,000 square miles and to have persisted for at least 6,500 years. Blocked by the gigantic ice dams, Lake Tight's water, and that of other finger lakes, rose higher and higher until it could surge over drainage divides like overflowing bathwater and create new channels. These sometimes ran in the opposite direction from before, and their massive runoff cut below the old Teays drainage level. This "Deep Stage" drainage heralded the beginning of the upstart Ohio River system, though more recent glaciers also influenced that river's modern course.

Today, one can read signs of this drama of ice and water throughout the hill country. The old Teays originated in North Carolina and likely crossed Ohio from south to northwest, probably exiting across Indiana and Illinois and emptying into a northern embayment of the Gulf of Mexico. If

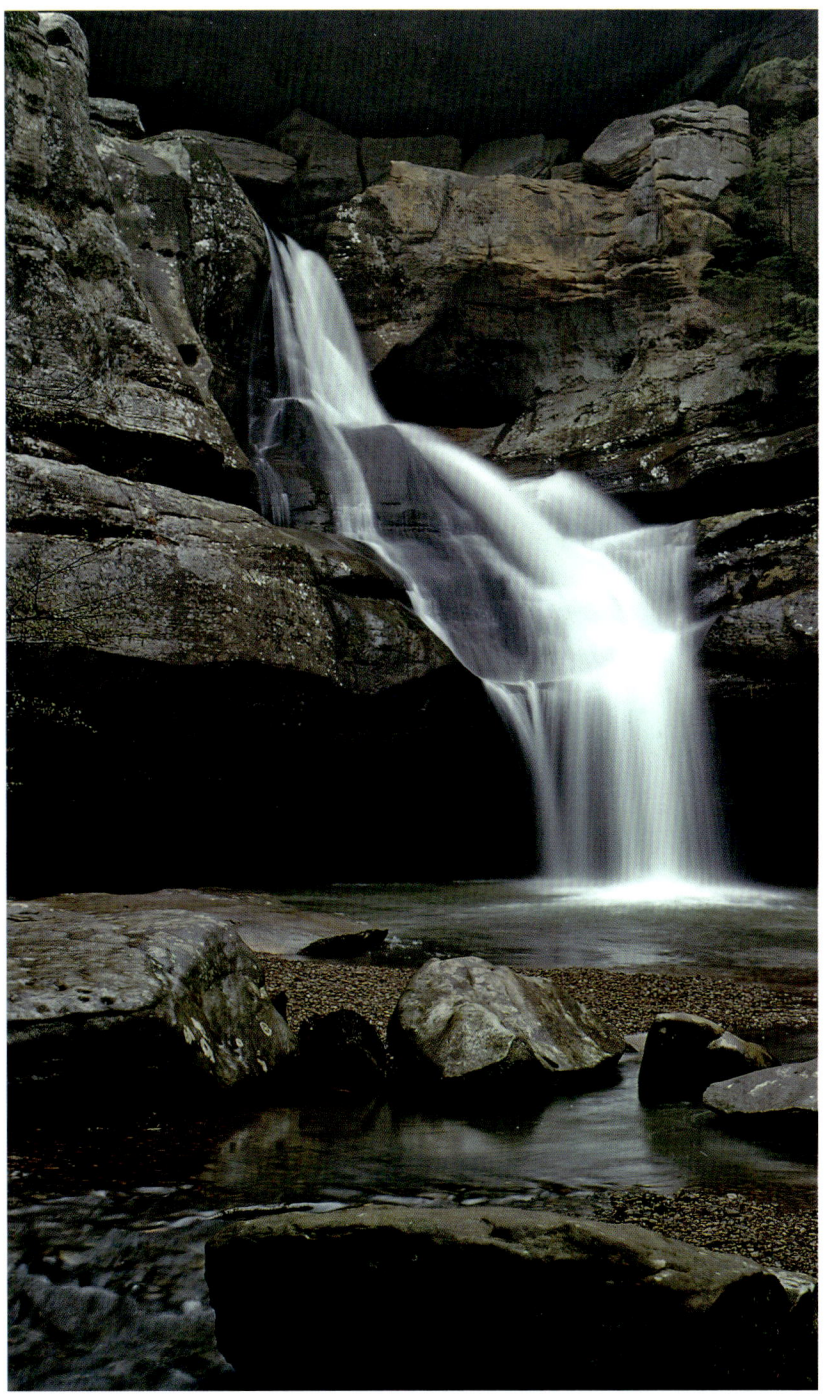

Cedar Falls rushes impressively through a bowl of Black Hand sandstone into its plunge pool beneath. Water action is eroding the soft middle layer of sandstone to form a rock overhang or "reentrant" behind the falls. Larger reentrants are found at Old Man's Cave, Cantwell Cliffs, and Ash Cave. During a dry summer, these falls, like many area streams, will dwindle to mere trickles. *Gary Meszaros*

you look at a satellite image of the Ohio Valley near Portsmouth, you will see clearly the large valley of a ghost river—the Teays—that curls from Wheelersburg to Waverly. At Waverly on the glacial margin, the Teays Valley dives beneath thick glacial till, and its course becomes invisible. From there north to Chillicothe, the opportunistic Scioto River runs in

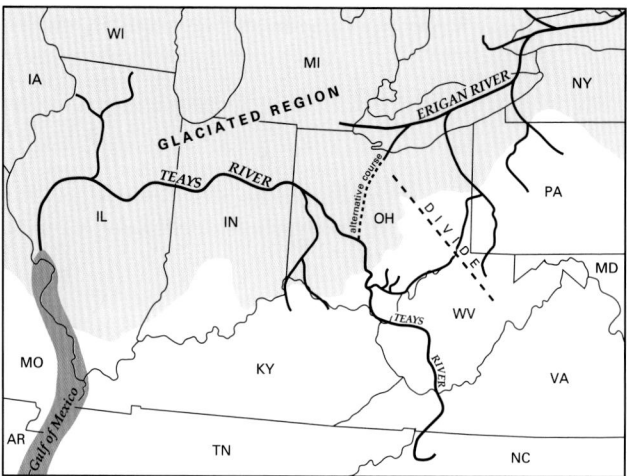

The preglacial Teays River valley, whose probable route from the Appalachians to the Gulf of Mexico appears here, once provided a corridor into Ohio for various southern plants. In rare hill country sites, some of these still survive, including flame azalea, rhododendron, bigleaf magnolia, and Canby's mountain lover. *Ohio Department of Natural Resources*

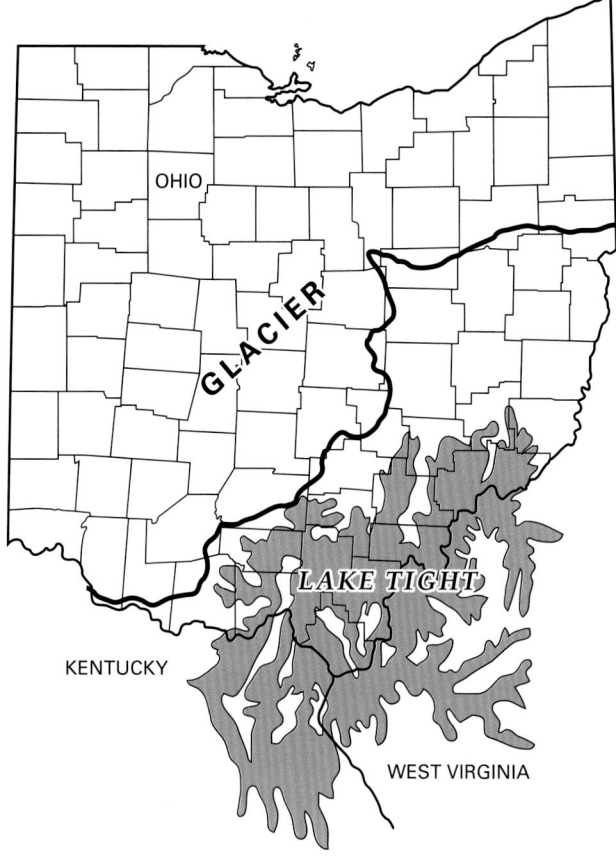

Dammed by an ice-age glacier, the ancient Teays River backed up into a huge system of finger lakes later named Lake Tight. Lake Tight may have looked something like the map above. This was the beginning of the end of the mighty Teays, replaced eventually by the modern Ohio River. *Ohio Department of Natural Resources*

the Teays' old valley—but toward the south rather than the north.

Many other modern streams occupy broad valleys of ancient rivers killed by the glaciers. The Muskingum, for example, flows in the old courses of two Teays River tributaries, again in the opposite direction from its predecessors, and even the Ohio River follows the Teays Valley for some distance. Glacial outwash covers many river terraces, deposited by phenomenal surges of meltwater as glaciers retreated northward. These floods also scoured many ravines steeper and deeper, as in Monroe County's "Little Switzerland" near the Ohio River. Captured tributaries often enter larger streams at odd, rearranged angles in a barbed pattern unlike ordinary dendritic, or tree-shaped, drainage patterns.

Plant Cover

During all this geologic commotion, the plant kingdom had not been idle. After the advent of flowering plants during the Mesozoic, the Age of Dinosaurs, many species migrated here from the southern Appalachians, probably along the ancient Teays and its tributaries. Eons before early European settlers arrived about 200 years ago, plants had wrapped the hills and ravines in a dense, luxuriant mantle; at the time of settlement, this included remnants of the ancient temperate forests that flourished here before the glaciers. Author Conrad Richter, whose trilogy of novels (*The Awakening Land*) traces settlement of the Ohio hills, begins his saga in *The Trees* as the Luckett family gazes out over the vast timberland that is about to become their home:

> For a moment Sayward reckoned that her father had fetched them unbeknownst to the Western ocean and what lay beneath was the late sun glittering on green-black water. Then she saw that what they looked down on was a dark, illimitable expanse of wilderness. It was a sea of solid treetops broken only by some gash where deep beneath the foliage an unknown stream made its way. As far as the eye could reach, this lonely forest sea rolled on and on till its faint blue billows broke against an incredibly distant horizon.

On the unglaciated Allegheny Plateau, varied oak and oak-hickory forests plus a diverse ancient forest community, named "mixed mesophytic" by the eminent Ohio botanist E. Lucy Braun in 1916, were the most pervasive when settlers arrived. Along the Ohio River in Lawrence and Scioto counties, in the Scioto River Valley, in parts of Hocking and Vinton counties, and in the old Teays River Valley of far southern Ohio are also remains of a third forest type, bottomland hardwoods. Composed of elms, sycamores, river birches, silver and red maples, and other water-tolerant trees, this association is least known, since river bottoms offer farmers some of the best agricultural land. Most bottomland hardwood forests long ago disappeared.

The trees were huge, especially the great hollow sycamores that grew in the river bottoms. Henry Howe, in his *Historical Collections of Ohio,* mentions one near Waverly in Pike County that was used as a blacksmith shop. In Scioto County grew another tree that could accommodate fifteen horsemen; the farmer eventually cut it down, says Howe, because his bulls were fighting and goring each other inside the giant trunk. A huge oak standing near Sarahsville in Noble County as late as 1880 measured thirty-four feet, six inches, just above the roots.

Using the encyclopedic observations of a lifetime's

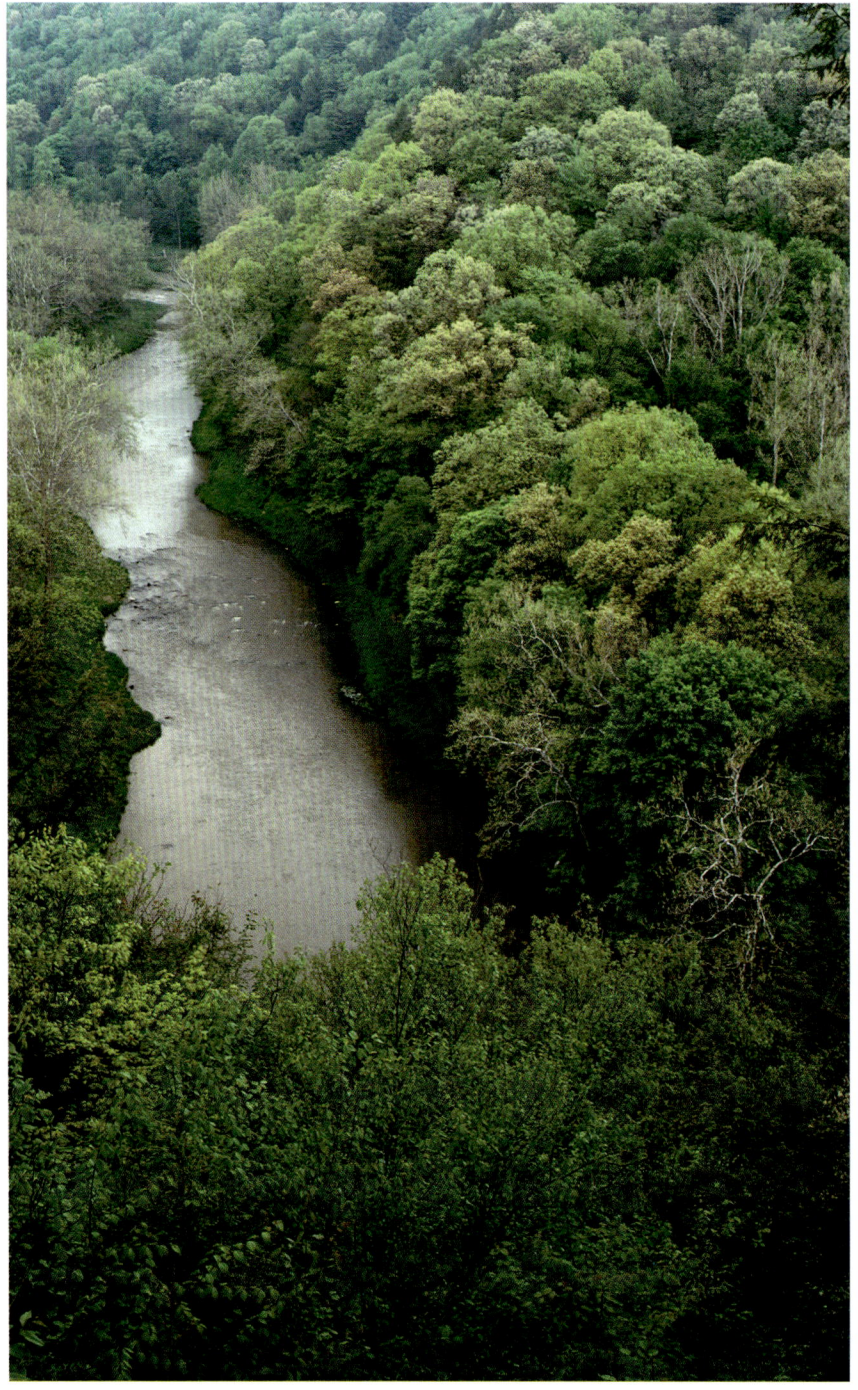

fieldwork, Braun argued that the mixed mesophytic forest is a direct descendant of the broad-leaved forests blanketing eastern North America before the glacial age. Its remnants now cover southeastern Ohio, western West Virginia, and parts of Kentucky and Tennessee, and the term that Braun originated to describe it is now commonly accepted. Usually found on fairly moist north- and northeast-facing slopes, this association's richest development in Ohio is probably in the forests of Jackson and Scioto counties.

The refrigerating effect of the glaciers, Braun believed, had not extended very far south of their leading edge, and the old deciduous forest had survived in the Cumberland Mountains and on the Cumberland and Allegheny plateaus in the southern Appalachians, including southeastern Ohio. From this center, the forest colonized barren land left by the glaciers as they disappeared northward. Species composition of these forest extensions and margins is less varied than in the mixed mesophytic forest itself, however. The reason for the difference, Braun explained in *Deciduous Forests of Eastern North America* and other writings, is that immature, less-dissected erosion surfaces on the glaciated land offer fewer specialized niches and microclimates for plants. The effects of a warmer, drier age persisted for some time after the glaciers melted and also influenced which plants of the old forest could survive and flourish on the new land. The beech-maple

Little Beaver Creek, seen here from above, cuts through Columbiana County in the northeastern corner of Ohio's hill country to join the Ohio River. Its unusually pure waters offer one of the few state habitats for hellbenders, large but harmless aquatic salamanders. *Gary Meszaros*

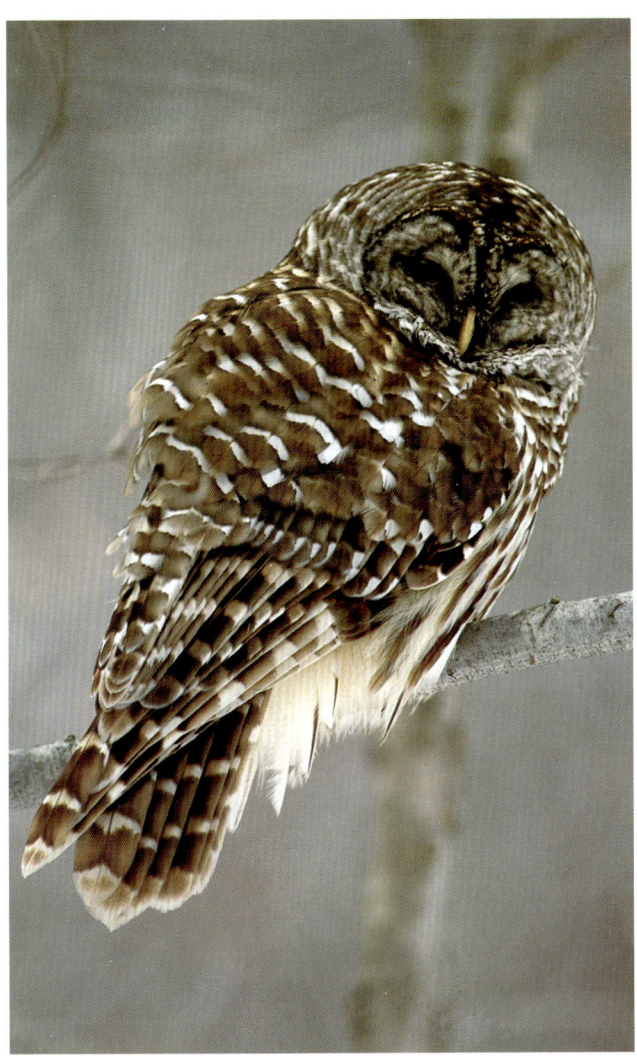

forests of northern and western Ohio and associations of oaks and hickories in the western counties are examples of these younger, less diverse woodlands. It is in the hill country that the richest forests remain.

The Hocking Hills area offers good examples of varied hill country woodlands. Because it lies in a transition zone between north and south and between the Allegheny Plateau and the Central Lowlands province to the west, the area is botanically quite diverse. A number of northern species are at their southern limits here, and southern species at their northern ones. Giant hemlocks and yellow birches line the deepest, shadiest ravines and north-facing slopes, considerably south of their usual

The barred owl, about three-fourths the size of the great horned owl, lost much of its wooded Ohio habitat due to agricultural clearing. Resurgent forests in the hill country have brought back its urgent question, "Who cooks for you?" Barred owls on occasion utter bloodcurdling shrieks, ow's, ah's, and guttural moans, enough to scare the daylights out of anyone hearing them in the woods. *Gary Meszaros*

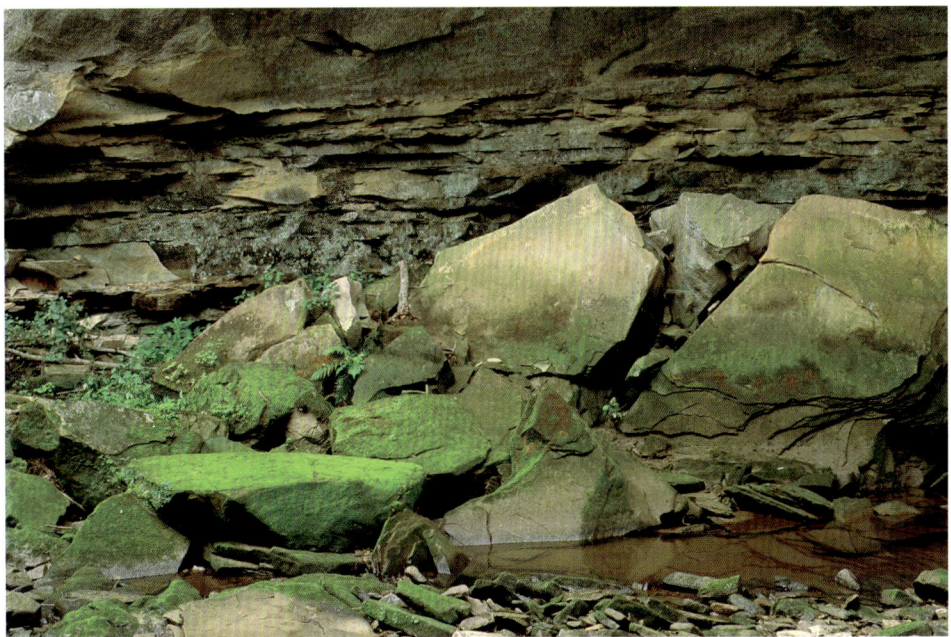

Softer layers of sandstone erode faster than more resistant ones do. Eventually, the harder rocks are undermined and plunge to the cliff's base. Some of these blocks can be truly monumental. *Gary Meszaros*

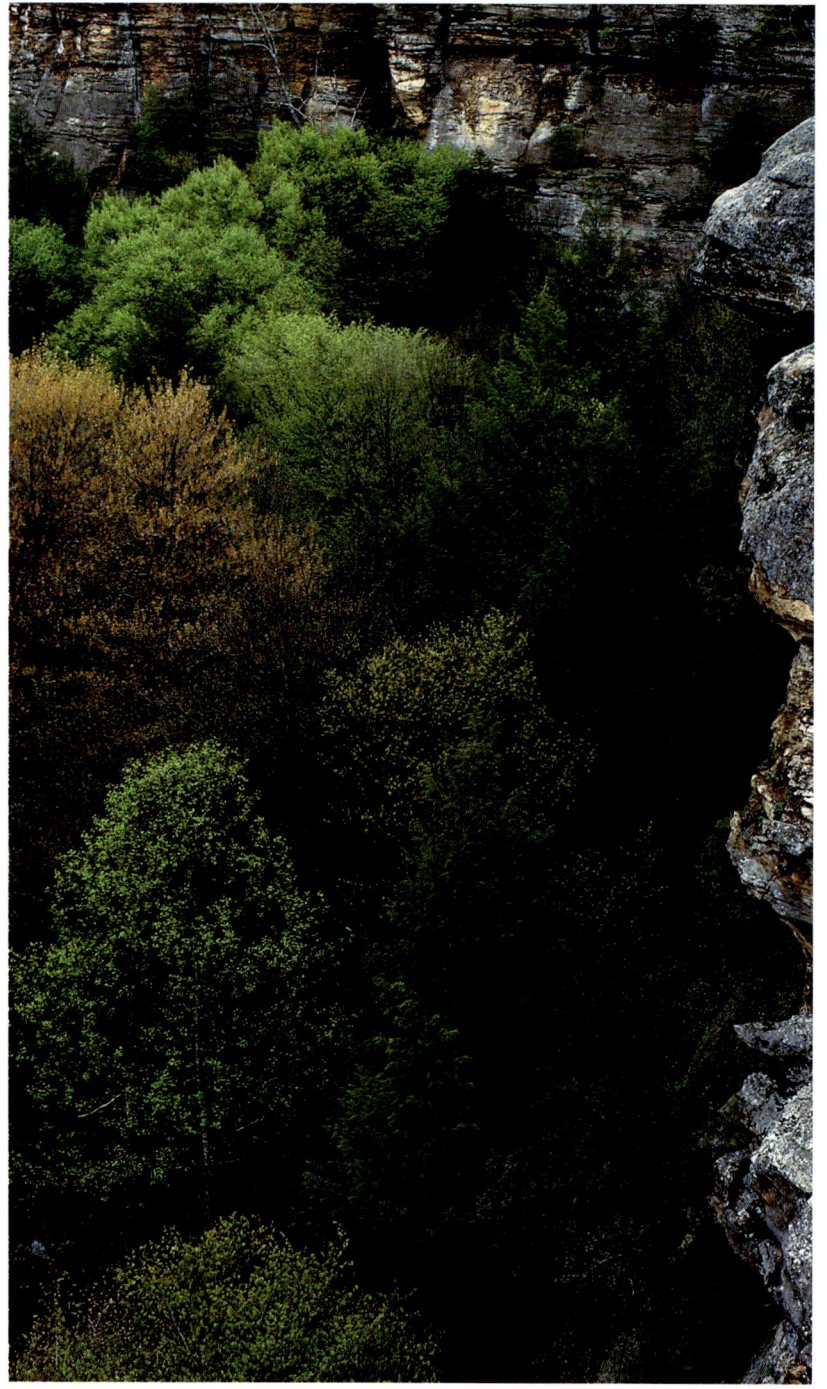

ranges. Settlers mistakenly named Hocking's Cedar Falls for the dark, graceful hemlocks that grew near it. Because steep gorge walls were inaccessible to loggers, some nearly untouched stands survived to be drawn into the new state forests and parks established during the 1920s.

On well-drained, more moderate sites grows Braun's mixed mesophytic forest. No species dominate this rich and varied grouping, which under good conditions can include 25 or more species in the canopy. Frequent members in the Hocking Hills region include black walnut, white basswood, bitternut and shagbark hickory, sugar and red maple, white ash, white and red oak, blackgum, wild black cherry, and sweet buckeye. Sensuously smooth gray trunks of beeches and the long vertical lines of tulip trees accent these woods, called "cove" forests by lumbermen because protecting cliff margins allow the timber to grow unusually tall, straight, and free from knots and wind damage. The woods also nurture a rich understory with many unusual southern plants, including lovely flame azaleas, rhododendrons, and sheets of spring wildflowers. Fine examples of cove hardwood forests grow at Rock House, Cantwell Cliffs, Conkle's Hollow, and other Hocking Hills locations.

Climbing up from the cool, dim ravine floors, you can feel the temperature rise as much as twenty degrees as you approach the harsher climate of the

Trails lead up the valley at Conkle's Hollow and also along the cliff rim, perhaps the most spectacular hike in the Hocking Hills area. The rim walk offers a splendid view of a fine cove hardwood forest clothed here in delicate spring greens. *Gary Meszaros*

ridgetops and hills. Cove forests generally grade through a mixed zone of tulip trees and oaks to other oaks and pines of the uplands. These trees adapt well to the dry, acidic soil that forms atop the Black Hand Sandstone, rock so porous that water tends to percolate down and away from the ridgetops. Chestnut oaks dominate the less extreme upland sites, accompanied by white and black oaks, red maple, hickory, and a few other species. Before the chestnut blight struck early in the last century, American chestnuts were common as well. This forest type is also found along the gorge tops of the Mohican River in Holmes County and Little Beaver Creek in Columbiana County, as well as on Carroll County's sandstone ridges, where chestnuts were once abundant.

Hocking's driest, poorest soil bears pines: pitch and scrub pines particularly, as do ridges in Jackson and Vinton counties to the south. Native pines also grow in southeastern counties along the Ohio River, on bluffs along the lower Scioto River, and at Black Hand Gorge in Licking County. Work crews planted large pine plantations in the Hocking Hills during the 1930s, 1940s, and 1950s, as well; these include American white and red pines, as well as Austrian pines and other European species. Similar plantations grow on slopes around Leesville, Clendening, Atwood, and Tappan reservoirs in the Muskingum Watershed Conservancy District where they stabilize eroded soil and help hardwoods grow again in cutover areas. Blueberries, huckleberries, trailing arbutus, and other members of the heath family grow low to the ground on Hocking's dry ridgetops.

Actually, the Hocking Hills area looks more as it did before settlement in the late 1700s than it did a

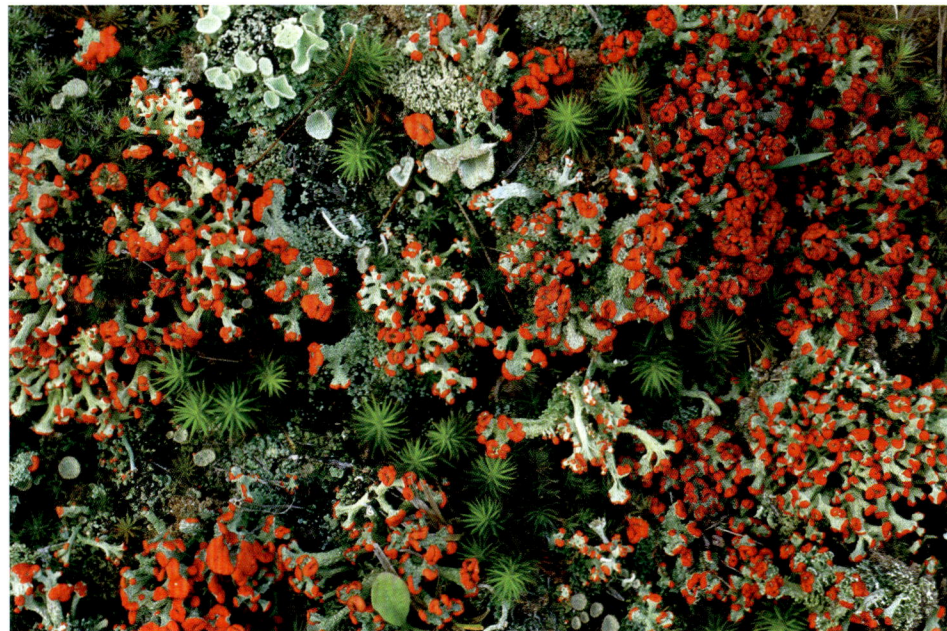

At the edges of many cliff tops lies a narrow strip of what can look superficially like northern lichen tundra. It has been exposed to the harshest wind action and the greatest temperature extremes in the area. These, for obvious reasons, are named British soldier lichens. *Gary Meszaros*

hundred years ago. In *A Botanical Survey of the Sugar Grove Region,* published in 1914, Robert Griggs lamented the ravages of portable-sawmill crews who bought clear-cut rights from local farmers, "cutting every stick capable of being made in a piece 2 by 4 inches or larger, with no regard for the future. . . . No scientific study is necessary to demonstrate that the land of this area is becoming poorer and poorer as its resources are dissipated under the present wasteful system of management. Everywhere one sees abandoned houses; in some parts of the area hardly half the houses are now occupied. Old, 'worn-out' fields are numerous and the number is increasing rapidly year by year."

The sandy soil, marginal even when prehistoric Indians hoed their small fields on the Hocking uplands long ago, simply could not support intensive and careless cultivation. By now, however, many of

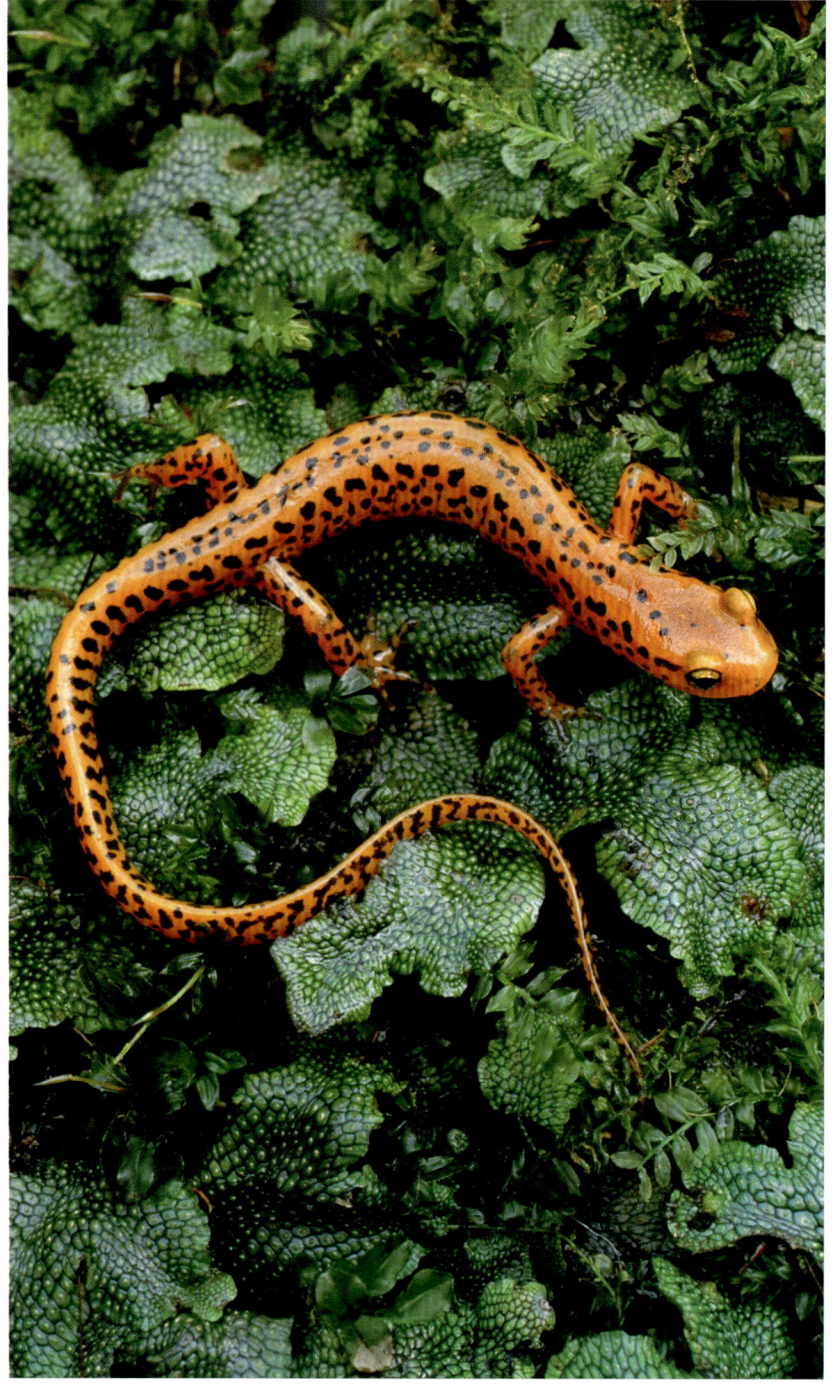

the abandoned farms have grown new forest cover, and the pine seedlings of midcentury are sizable today. Of course, the new forests are not botanically identical to the old ones, and many exhausted acres still produce only sumac, blackberries, and other scrub.

Many factors influence what kinds of woodlands grow in which places: bedrock, climate and moisture, soil types, altitude, disturbances of various kinds, direction of exposure, and so forth. Southeastern Ohio's sandstones and shales and lack of limy glacial till produce mostly acid soils, offering very different conditions to plants than do the "sweet" alkaline soils of western Ohio. Sandstone, too, is porous, which allows rainfall to percolate down through the rock and leaves higher areas in the hill country quite dry, hence the pines and "dry" oaks of the area's uplands. The deeply eroded landscape creates hot ridgetops, depressed frost pockets into which cold air sinks, and a variety of northern, southern, eastern, and western exposures.

Because the land's surface varies so widely in the hills, small microclimates support not only diverse plant associations but also a great variety of small birds, mammals, and amphibians, such as the many salamanders haunting springs that trickle from the porous sandstone where it overlies less porous shale. Spring birdsong still echoes the full-throated choruses of the past, and during the 1940s

The hill country's cool ravines and clear springs support a healthy assortment of salamanders. This long-tailed salamander, so called because its whiplike tail accounts for more than half of its body length, is one of Ohio's most beautiful amphibians. The green, scalelike growths on which it rests are liverworts, related to mosses. *Gary Meszaros*

biologists at the Ohio State University carried out a classic study of diverse microclimates in Hocking's Neotoma Valley.

Adams County on the edge of the Allegheny Plateau in far southern Ohio is an area of even more spectacular diversity than the Hocking Hills. Bordering the Ohio River about fifty miles east of Cincinnati, this county is Ohio's most sparsely populated and its poorest. There is little industry, and all over the forested hills of the eastern county, derelict barbed-wire fences mark the sites of defeated farms. As if to compensate for lack of human opportunity, the county boasts many rare animals and plants. It harbors the Allegheny wood rat, endangered in the state, and both the green and four-toed salamanders. Several very rare mollusks with fanciful common names like "three-horned warty-back" and "fragile heel-splitter" survive in the clear waters of Ohio Brush Creek and its tributaries. A number of reptiles and birds of special interest, including the lark sparrow (also declared endangered by the Ohio Department of Natural Resources), make the area their home. The hill country's edge is a botanist's paradise, the sort of place where a casually tossed pebble seems certain to graze at least one rare plant.

A unique set of geologic, climatic, and human influences have created this diversity. As already mentioned, Adams County straddles the boundary between the Allegheny Plateau and the Interior Low Plateau to the west; the Appalachian Escarpment runs north and south through the eastern third of the county above the valley that Ohio Brush Creek has carved on its way to the Ohio River. Here sprawls the most extensive privately owned group

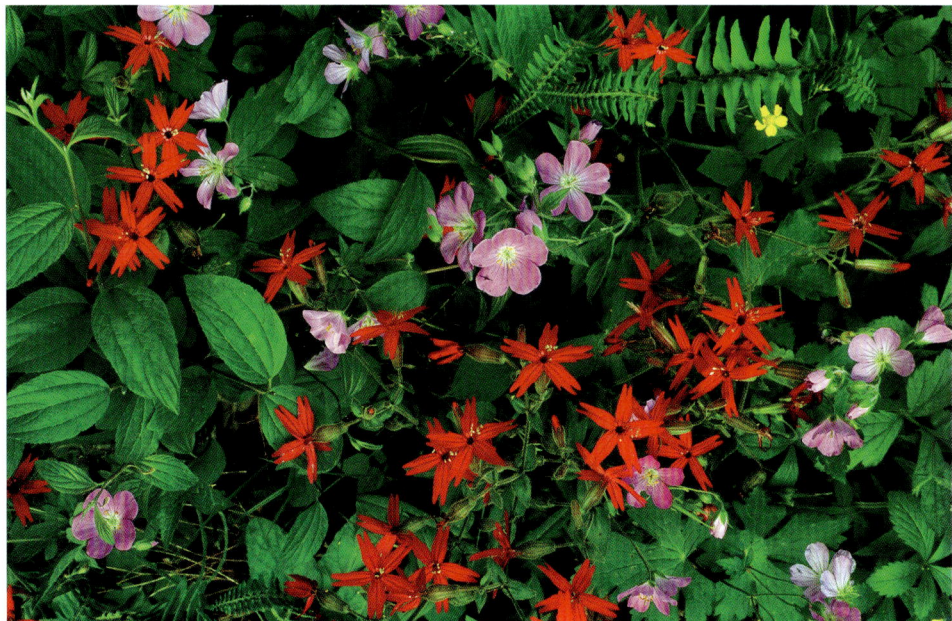

Blooming together on the forest floor, vivid fire pinks and wild geraniums light up late spring in the hill country.
Gary Meszaros

of nature preserves in Ohio, one of the largest in the country. This is the Richard and Lucille Durrell Edge of Appalachia Preserve System, managed by the Cincinnati Museum of Natural History. The preserves make up about 14,000 acres, protecting land that includes small prairies and a variety of forests, caves, stream riffles, and other habitats that support a prodigal variety of life on the "Edge." The area also includes several state nature preserves.

The first and perhaps most important cause of plant diversity here is the bedrock beneath. The area lies atop the eastern flank of the Cincinnati Arch. At the city of Cincinnati to the west, the rocks' dip is about twelve feet per mile, but it steepens to forty feet in Adams County. As a result, more layers can be seen in a shorter horizontal distance at the surface. The series of strata is further exposed by the Edge's vertical relief.

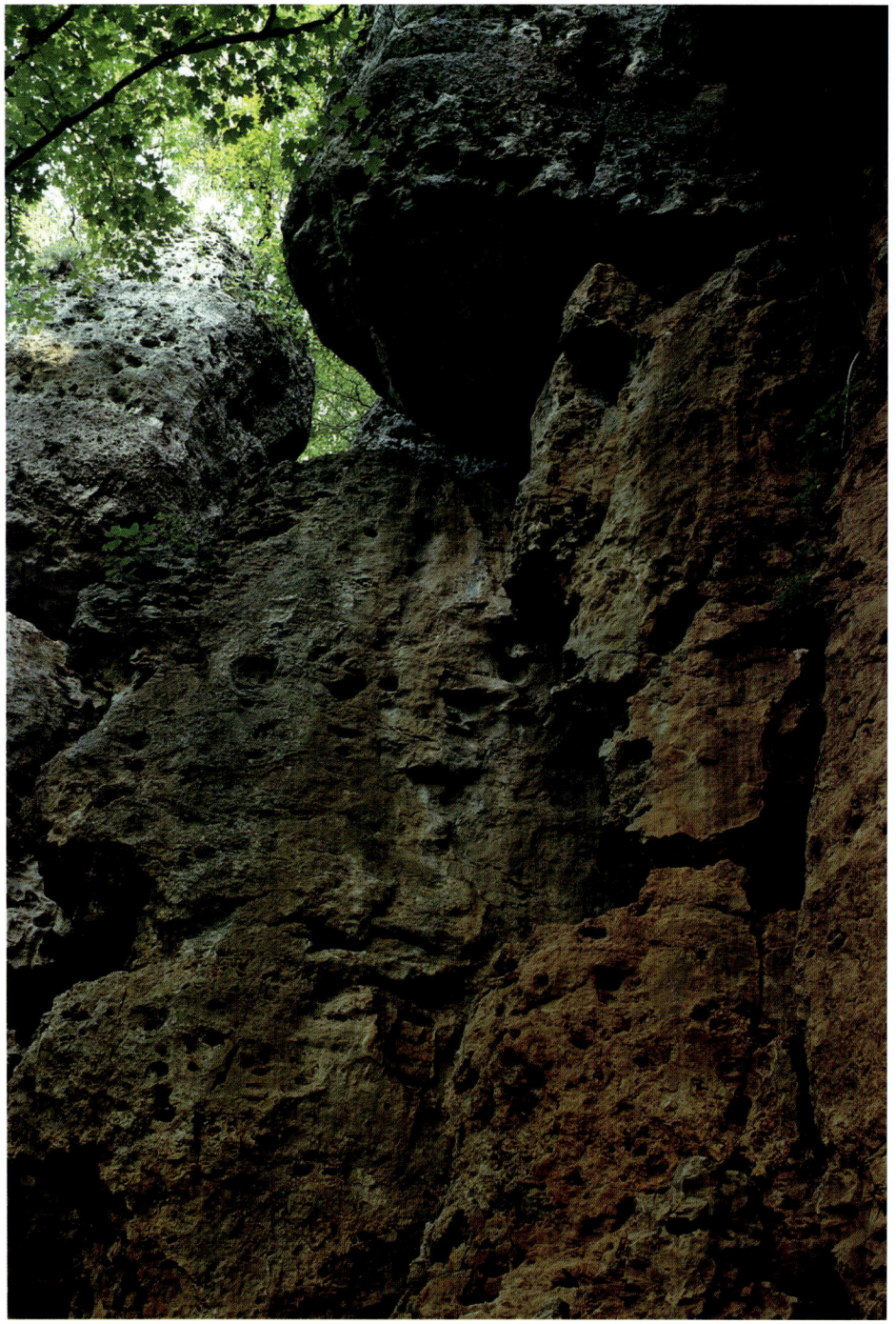

The important aspect for plant diversity is this: Silurian-age rocks in the county are, as we would expect in western Ohio, limestones, dolostones, and limy shales that form alkaline or "sweet" soils; the black Ohio Shale of Devonian age lying above them forms acidic or "sour" soils. And because of the rock layers' dip, bands of a variety of these bedrock-derived soils snake along the hills and ravines of the Edge. Plant differences sometimes render these bands visible from a distance along the hillsides: white serviceberry blossoms mark the boundary between the calcareous Peebles Dolomite and the Ohio Shale in spring, as do the subtly differing shades of young tree leaves on the various layers. In autumn the trees on the lower alkaline soils flame boldly against the more somber oaks of the acidic soils above. E. Lucy Braun observed that in Adams County, "The most striking correlation between vegetation and any environmental factor is the relation of plant communities to underlying rock."

A second reason for the county's rare species is the old course of the Teays River. While the river's main channel ran farther east, its tributaries drained this area. Along the Teays Valley, plants migrated from the southeastern Appalachians. A low-growing evergreen shrub named cliffgreen, or Canby's mountain lover, certainly one of Ohio's

Buzzardroost Rock is a huge cliff of Peebles Dolomite perched on the Appalachian Escarpment above the valley of Ohio Brush Creek. It weathers into a sort of honeycomb, hosting small rock gardens of columbine, purple-stemmed cliffbrake ferns, rock cress, and other plants, including the rare northern species Sullivantia. Atop the rock lies an ancient promontory prairie, which survives because its species are adapted to climatic extremes: hot sun, drying winds, and a water supply limited to immediate rainfall because of rapid runoff. *Gary Meszaros*

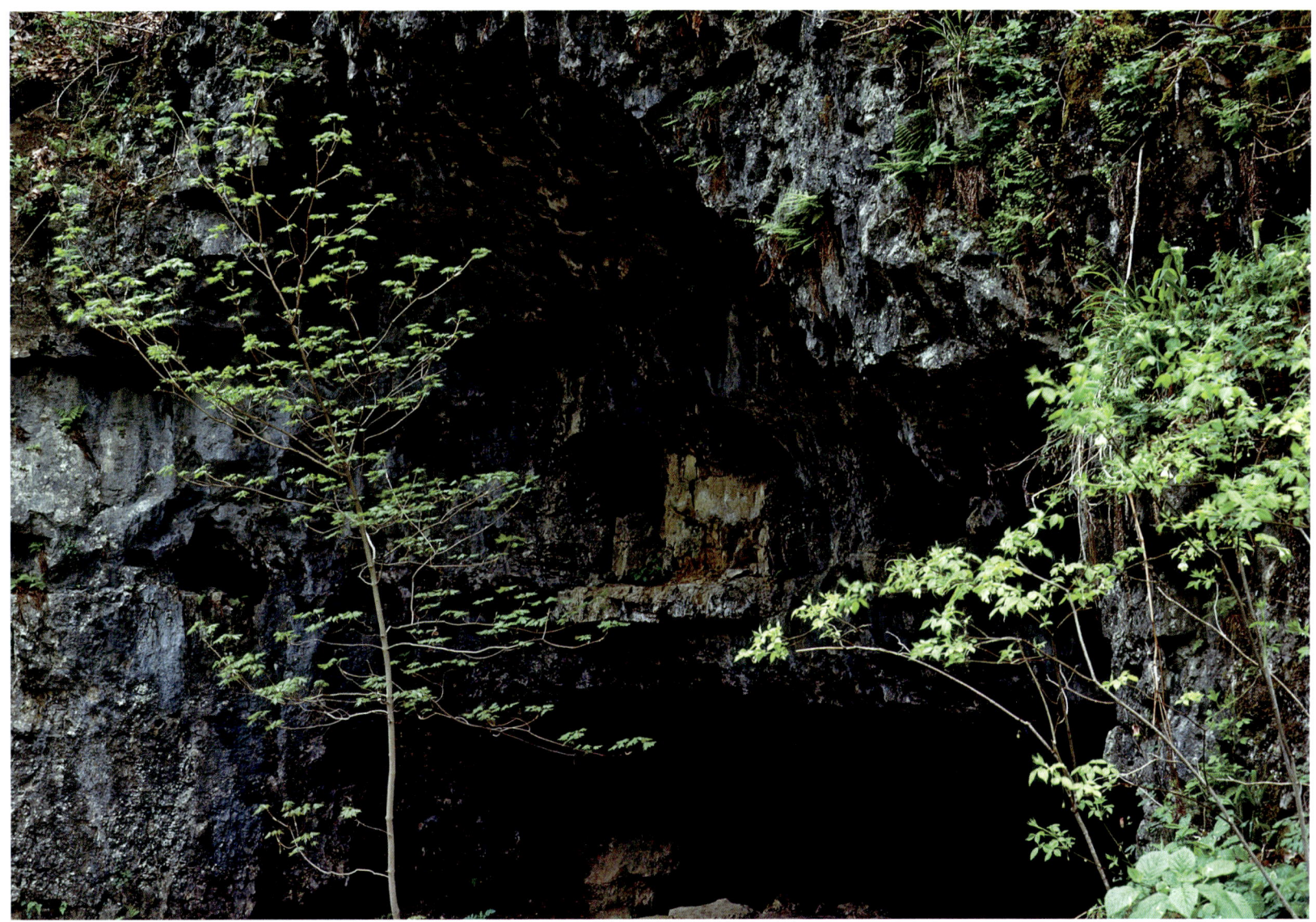

Just a few miles west of the glacial margin lie what were former-
ly known as Seven Caves, once a popular tourist attraction in
Highland County. They were probably formed as side galleries of
a larger cave whose back was broken by the great weight of an
Ice Age glacier. In 2005, the Arc of Appalachia Preserve system
at the edge of the hill country acquired Seven Caves and now
works to restore them as a natural habitat for cave creatures.
Gary Meszaros

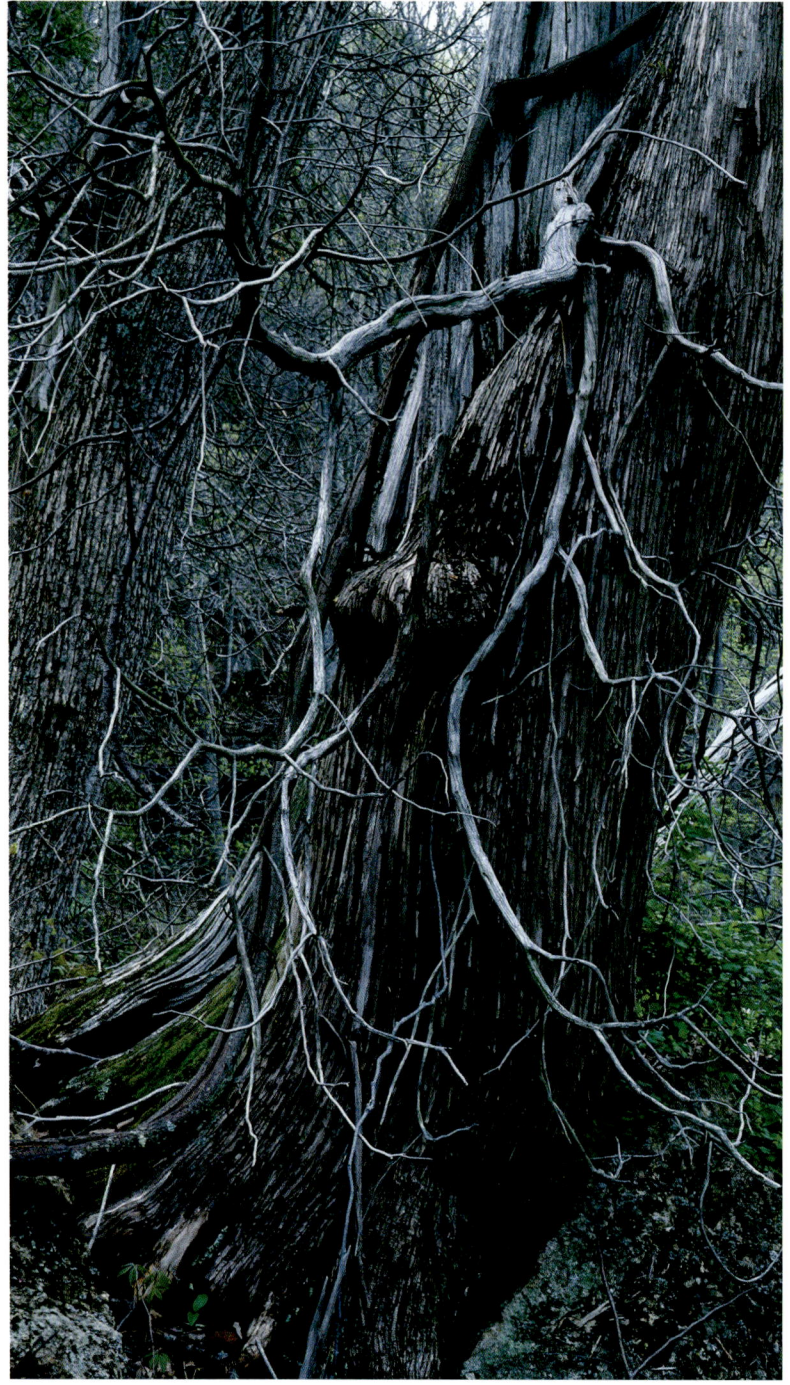

rarest plants, is perhaps one of these southern migrants. It grows on only two Ohio sites, one in the Edge preserve called the Wilderness and one at Fort Hill State Memorial in nearby Highland County. The glaciers' destruction of the Teays and the Ohio River system's birth left Canby's mountain lover and several other species stranded in southern Ohio, disjunct from others of their kind.

The glaciers themselves spared about two-thirds of the county, halting about ten miles northwest of the present Edge preserves. Thus, a roughly triangular piece of the Interior Low Plateau at the Appalachian Escarpment's foot is unglaciated—the only part of Ohio where Silurian limestones and dolomites lie unclothed by glacial rocks and gravels. Rising above this to the east, the Edge's deep hollows, rock shelves, and steep slopes and knobs have never been buried by glacial deposits, thus preserving many soils and microclimates to suit many different plants and animals, as elsewhere in the hill country.

Though the continental glaciers never covered this area, they did push before them populations of northern plants seldom encountered so far south today. Some of these species, notably northern white cedar (or arbor vitae) and the rare cliff-dwelling plant Sullivantia, survive in the cool gorges of Cedar Run and other area creeks and on dolomite cliffs unsuitable for the deciduous trees now dominant in the region. White cedar grows elsewhere in Ohio only at

The trunk of an old-growth northern white cedar stands near Cedar Run in Adams County on the edge of Appalachia. This species received its other common name, *arbor vitae* ("tree of life"), because it can live to be three hundred or more years old. Many of the oldest white cedars disappeared during the last century, logged to make lead pencils. *Gary Meszaros*

Clifton Gorge, Cedar Bog, and a few other sites. Because of the glaciers' influence, these unusual northern species joined Canby's mountain lover and other southeastern immigrants in this small area.

Another important climatic episode encouraged further plant groups to migrate into Adams County. As the ice sheets retreated, the climate warmed and became dryer, and for a time, about 4,000 to 6,000 years ago, climates favored xeric or drought-resistant species. This long dry spell formed what botanists call the Prairie Peninsula, an eastern extension of the great tallgrass prairies of the Mississippi Valley. Prairie grasses such as big bluestem, little bluestem, sideoats grama grass, and Indian grass colonized Ohio, accompanied by purple blazing stars, golden prairie dock, scarlet Indian paintbrush, and other brightly colored forbs—prairie plants other than grasses.

The more than 115 prairies remaining today in the Edge of Appalachia preserves are vest-pocket size, but they host very complex plant communities with a wealth of rare species. Certain southwestern plants, such as false aloe, also invaded the area from that direction. More recently, as the climate cooled and moistened, eastern deciduous forests pushed westward again, invading and obliterating Ohio prairies except for a few favored by local conditions. Thus, long before settlers saw the Adams County area, it included elements of three major

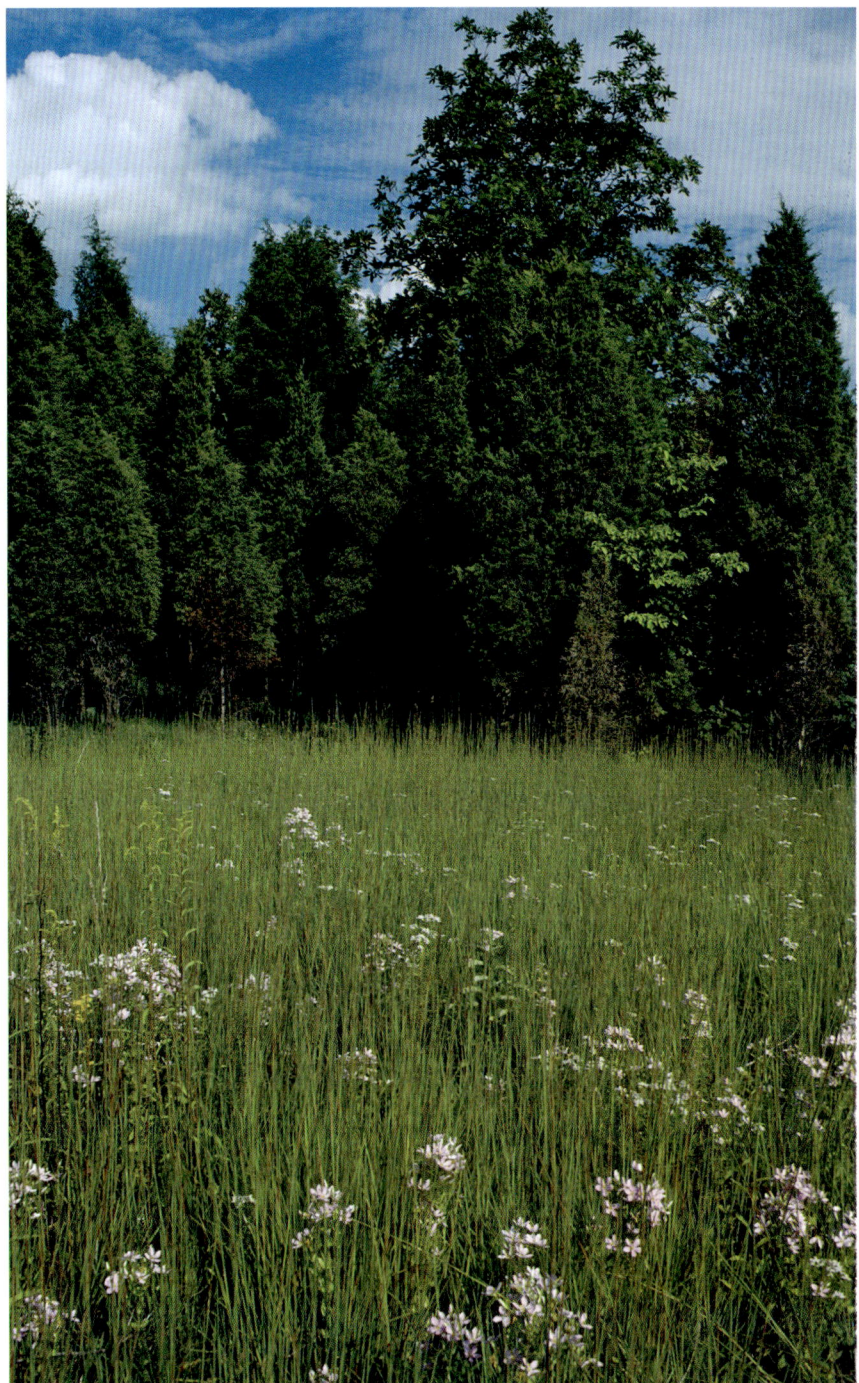

Lynx Prairie, one of the hundred-plus tiny prairies on the Edge of Appalachia in Adams County, was first set aside by The Nature Conservancy in 1959. It now belongs to the extensive Edge of Appalachia Preserve System. As in so many of the Edge system's remarkable 14,000 acres, it seems difficult to throw a stone without hitting a rare or endangered plant species. *Gary Meszaros*

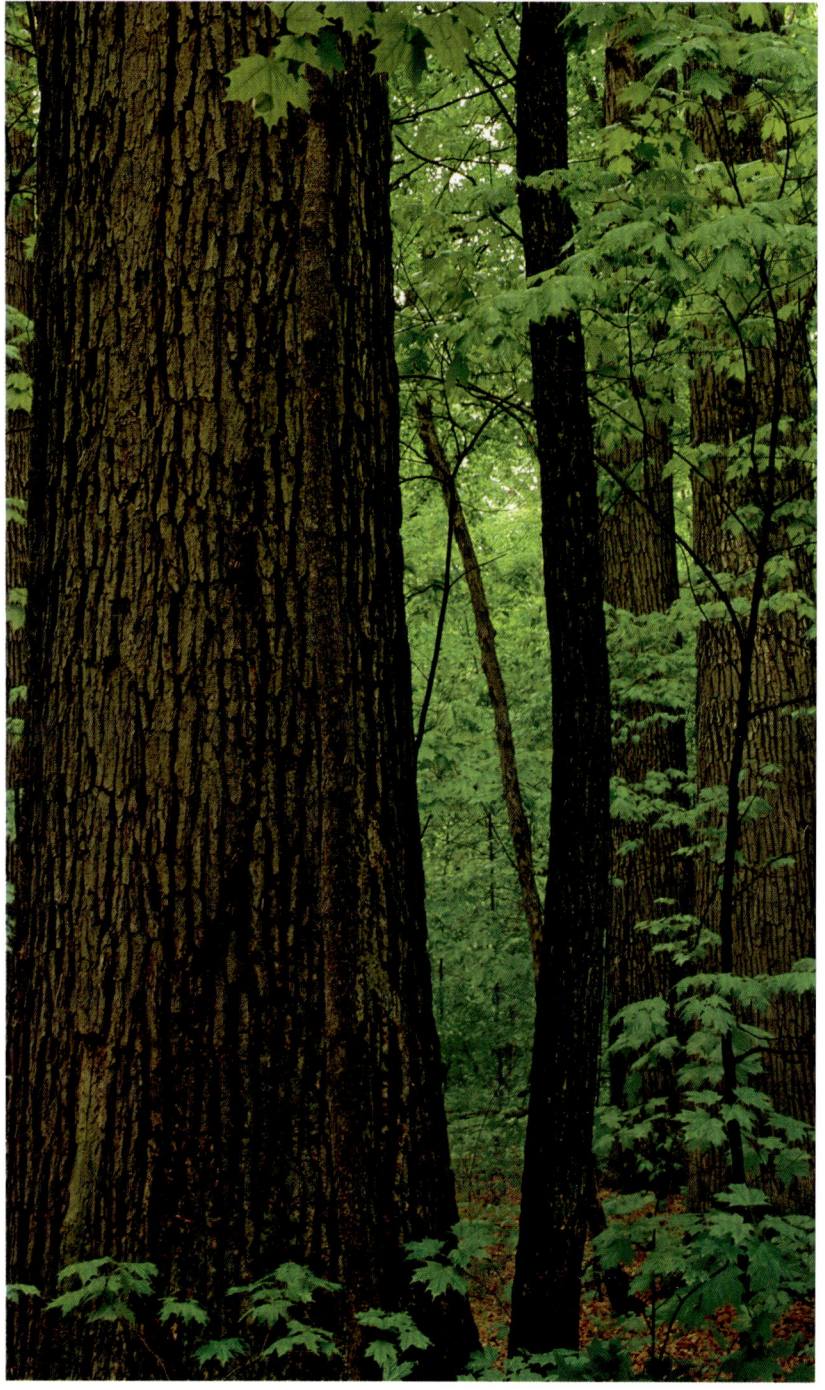

ecological communities, or biomes, that usually exist hundreds of miles from each other: eastern deciduous forest, tallgrass prairie, and northern hardwood or boreal forest. What an extraordinary mixture!

Both Hocking Hills and Adams County hint of the incredible richness and diversity encountered by European settlers of the Ohio Country, though the forests of neither area are old growth. Today, ancient forest survives in this state only in small, scattered pieces. A good way to view samples of Ohio's different forest types is to visit some of the Ohio Department of Natural Resources' forest preserves. These include, among others, Hueston Woods (Butler and Preble counties), Fowler Woods (Richland County), Kyle Woods (Mahoning County), Desonier State Nature Preserve (Athens County), Goll Woods (Fulton County), Gross Woods (Shelby County), Knox Woods (Knox County), and Gahanna Woods (Franklin County). In the hill country, the fifty-acre tract of primeval oak forest at Dysart Woods in Belmont County is the largest known remnant of the mixed oak woodlands that once covered so much of southeastern Ohio. Acquired by The Nature Conservancy, the woods were turned over to Ohio University to preserve and use as an outdoor laboratory. If you lean against a ponderous white oak trunk there and gaze into the high crown swaying slightly in the breeze, you will feel power and personality—it is easy to believe that the old tree is as aware of you as you are of it.

Old-growth forests linger only in small scraps here in Ohio. Most are preserved in Ohio Department of Natural Resources forest preserves. The largest remnant of the state's mixed-oak woodlands is Dysart Woods in Belmont County. *Gary Meszaros*

Settlers and the Forests

Some early residents found the big trees awe inspiring. Morris Schaff, writing about building the National Road through Licking County in "The National Road and the United States Road," recalled that when he was a boy in the mid-nineteenth century, "Three fourths of Etna Township was covered by a noble primeval forest. And now, as I recall the stately grandeur of the red oaks and white oaks, many of them six feet and more in diameter, towering up royally fifty and sixty feet without a limb; the shellbark hickories and the glowing maples, both with tops far aloft; the mold- and moss-covered ash trees, walnuts and the ghostly-robed sycamores, huge in limb and body, along the creek bottoms, I consider it fortunate that I was reared among them and walked beneath them."

Much more typical of early attitudes toward the forest, however, are those voiced by Sayward Luckett, protagonist in Conrad Richter's trilogy, *The Awakening Land.* In *The Trees,* she looks at the felled trees in her cabin clearing and thinks, "Those limbs a drying and leaves a dying would block out no more sun from her or her ground. They made a pleasing sight to a settler; for the best way such liked the trees was down, with their arms slashed off and ready for burning. The sweetest sound to a human deep in these woods was the hard whack of the axe, cutting or splitting, trimming or hewing, ringing a long ways through the timber till all the trees around knew what was coming to them."

Such attitudes toward the forest had a long history. Since before A.D. 800, Europeans had viewed the continent's wide forests as alien to humans—places of horror and bewilderment. The early Christian church advocated clearing this abode of the devil and of natural sin, and many immigrants brought the same attitudes with them to the New World hundreds of years later. The forest's beauty left most pioneers unmoved: As Alexis de Tocqueville noted when describing frontier life in the Michigan forests in *Democracy in America,* the pioneer "living in the woods . . . only prizes the works of man. He will gladly send you off to see a road, a bridge, or a fine village. But that one should appreciate great trees and the beauties of solitude, that possibility completely passes him by."

To settlers, the woods seemed endless, dark, and chaotic; they were physically trying and dangerous as well. People today sometimes need to be reminded how much less uplifting old-growth forests would be if shared with stock-harrying panthers, wolves, and bears, not to mention gales that could send enormous old trunks and branches crashing down on one's head. Early settlers also feared

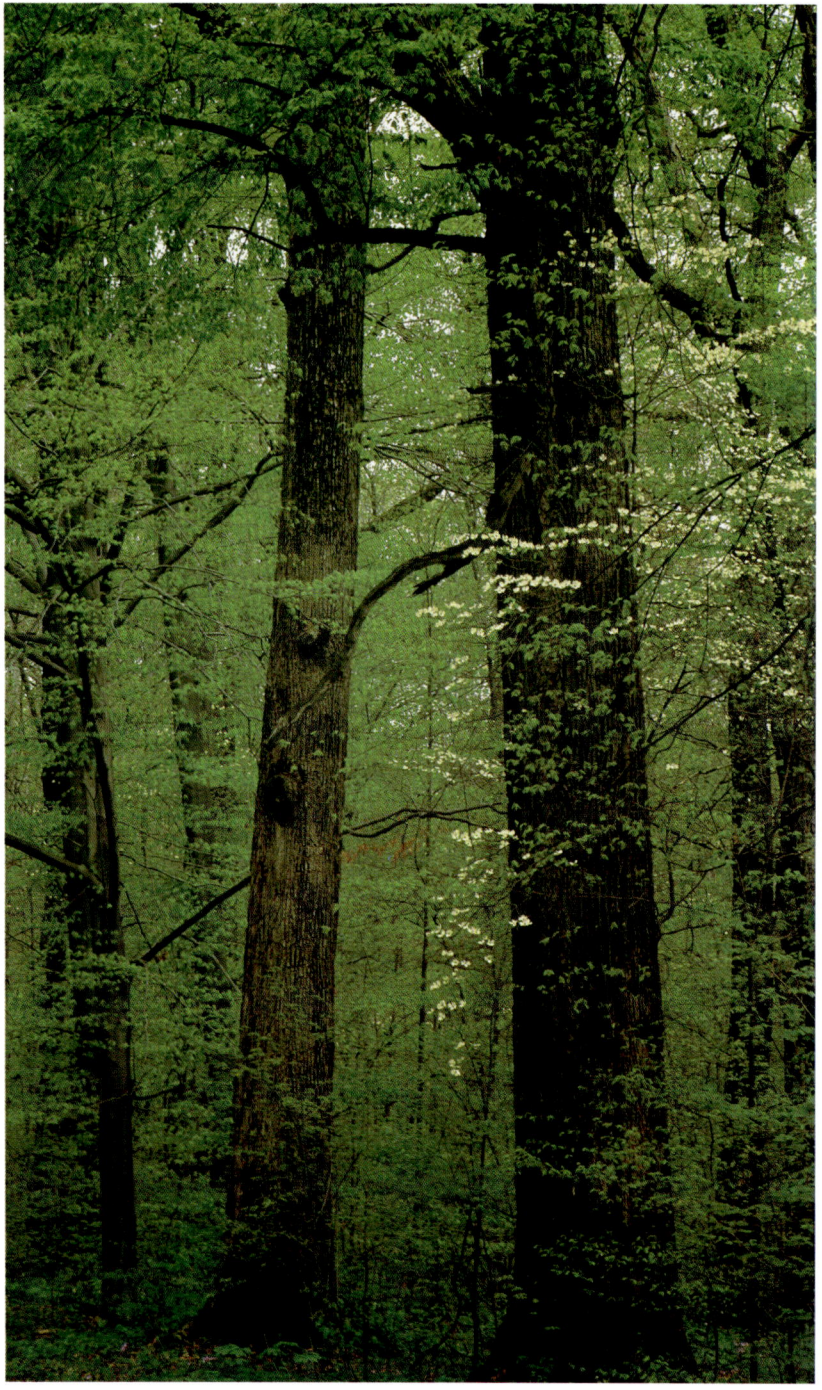

Indians who might be watching from the woods. Pioneers attacked the trees with an almost religious fervor, but their motivation was supremely practical as well: ideas of progress and development were natural outcomes of the struggle for survival.

During the two hundred years before 1850, settlers cleared—improved, they called it—100 million acres at the very least, 9.8 million of them in Ohio. Between 1850 and 1859, when clearing accelerated dramatically, almost 40 million acres fell, including another 2.8 million in this state. The open fireplaces of the day consumed vast quantities of wood, an estimated 268 million cords nationally between 1800 and 1809 alone. Settlers could also make a little ready cash by burning downed wood for potash and pearlash. These, in turn, produced lye, used in soap- and glass-making, tanning, bleaching, and various other processes.

In many cases, huge windrows of trees were simply burned, with no attempt made to use them for anything, and the cleared landscape was for years suggestive of untidy waste. Visiting Europeans deplored the untidiness more than they did the waste. Captain Basil Hall of the Royal Navy sketched a scene of newly cleared land and captioned it as follows: "The trees are cut over at the height of three or four feet from the ground and the stumps are left for many years until the roots rot;—the edge of the forest, opened for the first time to the light of the sun looks cold and raw;—the ground rugged and

Approaching the trunk of one of these ancient oaks can remind a walker of the staggering labor needed to clear off "big butts" with only crude hand tools and fire. For many settlers the struggle seemed truly a war with the trees. *Gary Meszaros*

ill-dressed . . . as if nothing could ever be made to spring from it."

Though settlers cleared the forests out of need, they also destroyed plants and animals on a devastating scale. Economically useless initially, the piles of felled trees burned, clouding the vast frontier sky with thick, blue haze. Then a nearly insatiable market developed. From 1826 to the turn of the next century, iron furnaces gobbled wood in southern Ohio's six-county Hanging Rock Iron Region, which also extended south of the Ohio River into Kentucky. By the 1860s, the state's railroads were burning a million cords of wood a year, as well as consuming many, many wooden railroad ties. Commercial lumbering moved into Ohio in the latter half of the century as well, and old farmers must have remembered with regret the prime hardwoods squandered on the waste piles of their youth. In the end, old-growth stands survived only on land too steep or inaccessible to log.

A Long Destruction

It is clear that farming was the greatest drain on Ohio's timber. However, in parts of the hills, especially in the Hanging Rock Iron Region and in counties along the Ohio River to the east, charcoal-fired iron furnaces accounted for the most ravenous devouring of forests. Furnaces of the Hanging Rock region centered in Scioto, Lawrence, Jackson, and Vinton counties, with a few more in Hocking and Gallia. Many of them have disappeared completely today, but ruined stacks can be seen at some, and a few stacks have been preserved, though the wooden buildings that surrounded them are gone. Many visitors have seen the old stack of Hope Furnace at Lake Hope State Park in Vinton County, and the Ohio Historical Society has restored Buckeye Furnace and its buildings near Wellston in Jackson County. Buckeye Furnace is the place to witness how a typical charcoal furnace operated.

This was a major regional business, with 46 furnaces firing at its peak in the mid-nineteenth century, with others in northern Kentucky. Charcoal furnaces operated in the Ohio section from 1826 to 1916, when Jefferson Furnace in Jackson County made its last blast—a total of nearly a century. Deposits of iron ore that were washed into the great Pennsylvanian-age swamps by rivers draining the Appalachians made this cluster of furnaces possible. Siderite (iron carbonate) and limonite (hydrous iron oxide) accumulated in the ancient wetlands above a limestone deposit named the Vanport, which the iron masters used as flux to remove impurities from molten ore. Add vast forests for charcoal, sandstone for building furnace stacks, cheap labor, and transportation via the Ohio River, canals, and later railroads, and you had an industry. The iron was high silica and high quality. England used it to wage the Crimean War; a few years later demand was so great that iron masters sometimes shipped pig iron to the Civil War so hot from the furnace that it burned the teamsters' wagons.

Operators needed immense quantities of raw materials. One ton of pig iron consumed from 150 to 200 bushels of charcoal, 5,000 pounds of ore—mined from beds only about a foot thick—and about 300 pounds of limestone. Yearly production averaged about 2,000 to 3,000 tons per furnace. Each of the 46 monsters consumed from 300 to 350 acres of wood in a single year. During the life of the Hanging Rock industry, a majority of forests in the region were cut several times at intervals of 20 to 30 years. In *Historical Collections of Ohio,* Henry Howe noted that by the mid-1840s, "In the winter season about 500 men come from abroad, to cut wood for the furnaces in Lawrence [County], some

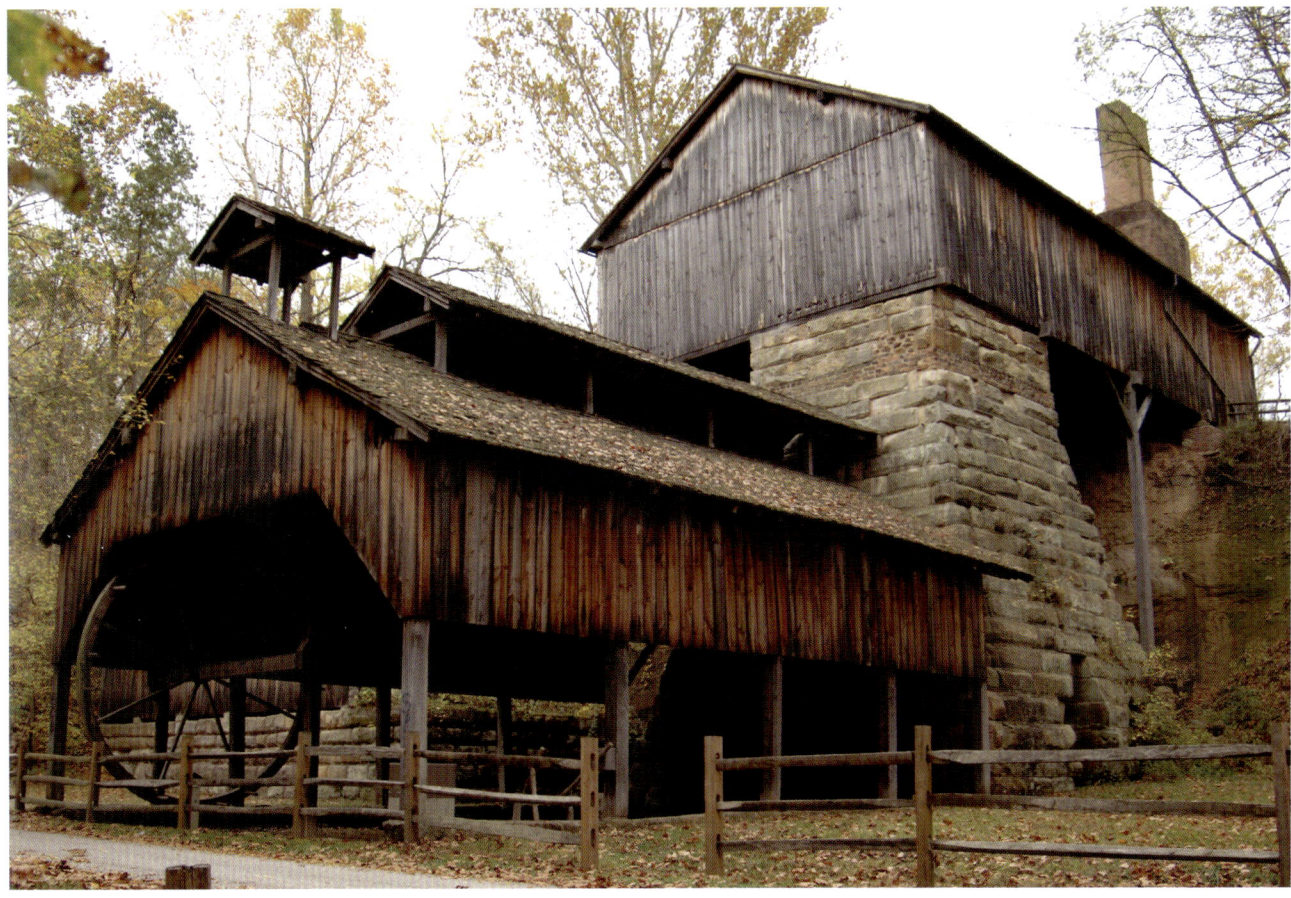

Buckeye Furnace, built in 1851 in eastern Jackson County, produced about 12 tons of pig iron a day until its final blast during 1894–95. In 1976, the Ohio Historical Society reconstructed its buildings as a state memorial open to visitors. The old stone stacks of a number of other furnaces still survive in the hill country as well. *Ohio Historical Society*

of whom walk distances of hundreds of miles from their cabin homes among the mountains of Virginia and Kentucky." He also noted that the furnaces and their hundreds of workers were an important market for farmers, which must have caused even more forest clearing.

Dwindling wood supplies contributed to the industry's decline, along with competition from higher-quality iron ore discovered in northern Minnesota later in the century. New furnaces fired by stone coal and coke also hastened charcoal fur-

naces' end. The nosedive began after the Civil War, accelerated in the 1880s, and finally ended in 1916 with the last blast at Jefferson Furnace. By that time, today's rust-belt cities of northern Ohio and western Pennsylvania had supplanted the old iron belt of Hanging Rock, and the forests began to regrow.

Not only did the woodlands suffer but, inevitably, so did the animals that lived in them. The nearly complete destruction of Ohio's ancient forests went hand in hand with a frontier war against large predators, such as wolves, cougars, and bears, and

as wood bison, white-tailed deer, turkeys, and the soon-to-become-extinct passenger pigeon were put to use for food, hides, and trade. Authorities offered bounties for killing wolves and cougars and sometimes exacted penalties from those who did not account for their quota of dead gray squirrels, whose huge numbers made them a threat to crops.

Sometimes large-scale hunts were held, the most successful of these being the Great Hinckley Hunt of December 14, 1818, in northeastern Ohio. Six hundred men and boys carrying muskets, bayonets, butcher knives, and axes formed a cordon around the township's perimeter and drew the circle tight. Dogs then drove the animals caught inside it into the men's musket fire. At the end, hunters shot down into masses of panicked animals trapped in the bed of a frozen stream with steep, high banks. The men celebrated that night with barbecued bear meat and large quantities of whiskey. Their "war of extermination" and destruction of forests spelled

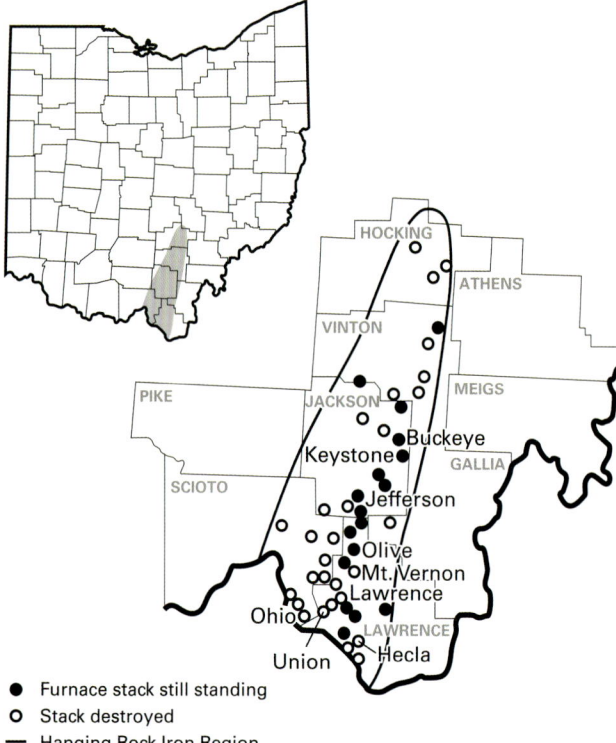

The Hanging Rock Iron Region extended from Greenup County, Kentucky, north to Hocking County, Ohio. It produced charcoal iron from 1818 to 1916 and was a major force in establishing Ohio as an industrial state. High-quality Ohio iron was reportedly used to construct the ironclad ship *Monitor* during the Civil War. *Ohio Department of Natural Resources*

the wholesale killing of other animals, as well. To be fair, cougars and bears could sometimes be dangerous, and wolves, though not threatening to human beings, had figured in so many lurid folktales over the centuries that they were universally hated and feared. All of these creatures sometimes took sheep, cows, pigs, and other livestock, especially as settlers disrupted their forest environments and killed off their natural prey. Other animals, such

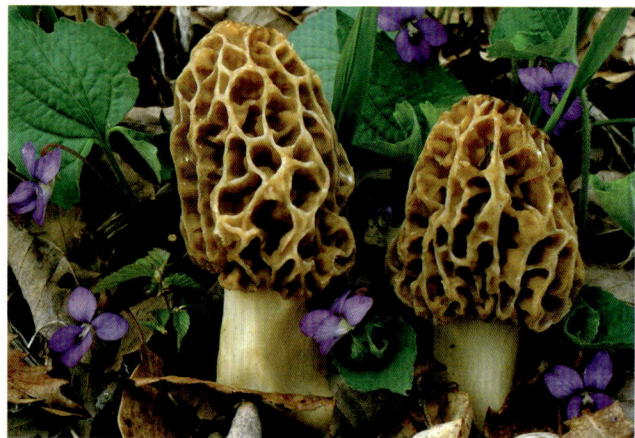

An early morel leads off a seasonal succession of many fungi. Nestled here among blue violets, it is eagerly sought by foragers for its nutty flavor. *Gary Meszaros*

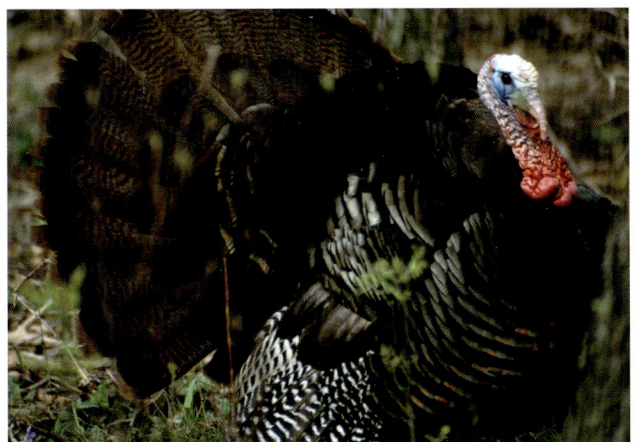

A wild turkey gobbler displays with great fanfare in spring. Turkeys often travel along a trail in single file, a trait that early Ohio hunters used to their advantage, awarding prizes to marksmen who downed the most birds with a single shot. Wild turkeys were once completely wiped out in Ohio, but, in recent decades, they have been successfully reintroduced in many places. *Gary Meszaros*

doom for the big carnivores; in 1818, the same year as the Hinckley hunt, Ohio discontinued its wolf bounty. In 1838, the pioneer naturalist Jared Kirtland observed that the cougar, or "mountain tiger," had disappeared from the state.

By 1830, beavers were also only a memory in the young state of Ohio. The fur trade, of course, was their nemesis, fueled by the European love of beaver hats. Trappers had taken them and other furbearers in great numbers here for two hundred years. Black bears, wood bison, and elk were also rare by 1800, and lynx, bobcats, martens, and fishers essentially vanished by midcentury. By around 1900, so did playful river otters. About the same time, wild turkeys and white-tailed deer were completely extirpated from the hill country and from the rest of the state as well. Early in the twentieth century, wood ducks were in danger of extinction,

and Canada geese were also in serious decline. Many smaller creatures, especially those associated with wetlands that were systematically drained and rivers that were dammed for mills and otherwise degraded, also declined or disappeared. Long-established ecosystems shattered, and many species would have suffered heavily even without active "wars of extermination."

Fortunately, nature is resilient, and with benign neglect or active stewardship, it does rebound. One can hardly overlook the fact that white-tailed deer and Canada geese have returned to the hills; beaver, wild turkey, river otters, and wood ducks have also made comebacks with help from state agencies and concerned individuals. Regrowth of timber has begun to attract black bears into the eastern counties from Pennsylvania and West Virginia, and if

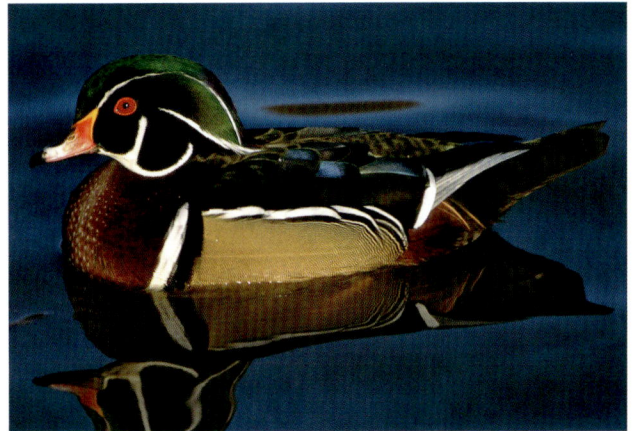

The male wood duck, with his bright, crisply defined markings, is among the most beautiful of ducks. It has been said that at the turn of the twentieth century there were more wood ducks in Belgium, where they were raised by bird fanciers, than in all of North America. Closely controlled hunting seasons, provision of artificial nest boxes, and the regrowth of forests that can provide nest holes for wood ducks have aided their comeback. *Gary Meszaros*

given a chance, they should build viable populations, especially in the hills. Coyotes have invaded each one of Ohio's 88 counties from westward. Of course, the futures of all these animals, as well as the ecological balances among them and with us, depend on our own actions: our caring for the hills' ecosystems, our refusing to degrade them further, as well as fostering their recoveries.

Other Rents in the Fabric

Today, uncontrolled growth, especially along interstates, as well as extractive industries like coal mining, are probably bigger threats to natural hill country ecosystems than is agriculture. Mark J. Camp comments in *The Roadside Geology of Ohio* about Belmont County, the all-time leader in Ohio coal production: "Commercial sprawl spreads along the routes near St. Clairsville, competing with mining at destroying the natural landscapes." Certainly, in this area I-70 can often look like a linear shopping mall.

The first recorded coal mining—in Jefferson County—dates back to 1800, though settlers attacked surface coal with pick and shovel to heat their homes even earlier than that. Commercial mines exploded along with railroads in the 1860s, as the iron horse developed a big appetite for coal, as well as offering mine owners improved transportation to markets. Several methods of underground mining preceded surface, or strip mining, which, though it began in the 1880s, did not become common until the mid-twentieth century.

Henry Howe reported of Perry County in *Historical Collections of Ohio*,

New Straitsville is in the heart of the richest coal-producing district west of Pennsylvania; it is only three miles over the high, steep hills to bustling Shawnee, with its mines and blast furnaces; southward are Gore, Carbon Hill, and finally Nelsonville, all strong mining towns of the Hocking Valley. A good deal of life is underground. When a stranger comes to Straitsville and beholds a few houses on half-a-dozen ridges and but two streets of consequence, he is scarcely ready to think that there is a population of nearly three thousand in the town, but if he went into many of the houses he would find them packed with people, and very often one roof shelters half-a-dozen families.

Straitsville and Shawnee were desperate places during the great strikes that prevailed in Hoadley's administration. . . .

Howe fails to mention that during one of these disputes of the 1880s, striking miners pushed burning coal cars into a mine owned by the New Straitsville Mining Company, setting the mine afire. They put wood in coal cars, soaked it with oil, and lit it. Then they pushed the cars into the mine, where the fire spread to the coal seam. The mine closed, but the New Straitsville mine fire still smoulders today—125 years later! More than two hundred square miles of coal is thought to have burned.

Since 1800, miners have excavated close to 4 billion tons from Ohio's coal measures. It's estimated

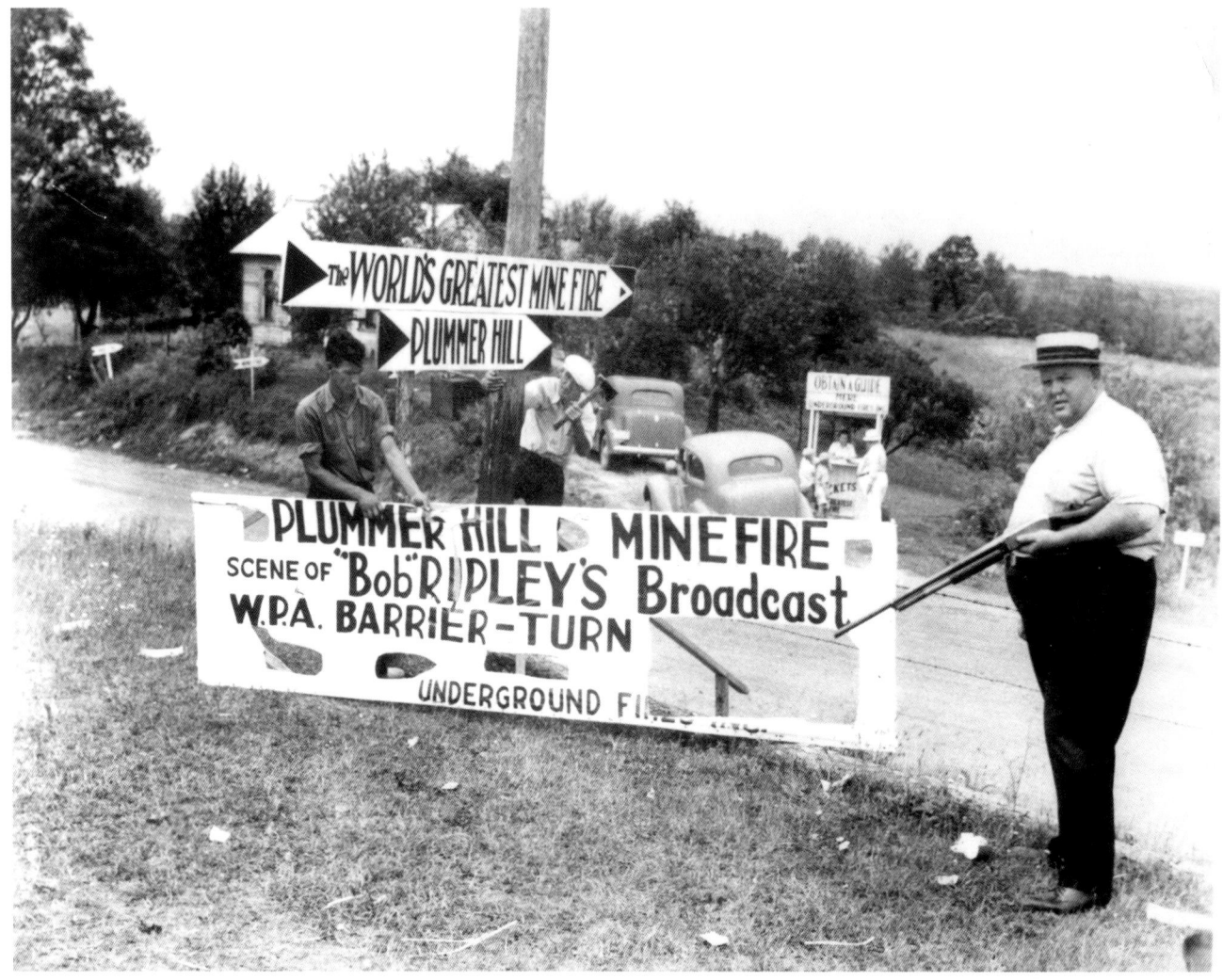

that they have mined more than 600,000 acres in the eastern part of this state. Until the 1940s, miners took almost all of their "black gold" from several types of underground mines. There are an estimated 6,600 of these old mines, and no maps exist for at least 2,300 of them. Unexpected cave-ins are a major problem, especially for homeowners in some communities, as are acids—known as "yellow boy"—that leach from old mines and kill life in rivers and creeks. Strip mines, which became common with advances in earthmoving technology during the 1940s, have caused other serious problems. Giant earthmovers, called walking drag lines, strip the earth and rock layers lying above coal seams, which naturally removes all vegetation, as well. Although strip mining is safer and more efficient in recovering coal than underground mining, its cost is destruction of the natural landscape. It also employs far fewer miners.

The most gigantic walking drag line, called "Big Muskie," was owned by American Electric Power and operated from 1969 to 1991 at its Muskingum Mine southwest of Cumberland in Guernsey County. Muskie was six stories tall and required the equivalent of an eight-lane highway for transport. A staggering 120,000 acres were stripped at this mine, until 1990s Clean Air Amendments and changes in the mining industry decreased demand for Ohio's high-sulfur coal. However, depletion of other fossil energy sources may again speed removal of this state's coal, 90 percent of which goes to produce electric power. New technology that promises cleaner coal burning also may result in expanded use. Whether it will solve problems of air pollution and climate change remains to be seen. Certainly, increased demand will raise pressure for expanded surface mining and for high-disturbance underground techniques such as long-wall mining; the latter has recently threatened that grand old oak forest preserve, Dysart Woods.

Meanwhile, American Electric Power (AEP) has worked to reclaim 30,000 acres of forest and 26,000 of grassland for public recreation, including 10,000 acres donated to make up the Wilds, a rare animal preserve and breeding center west of Cumberland. This is laudable, but no one should think that reclamation means restoration of land and ecosystems to their original states. The ReCreation Lands, as AEP calls them, are a peculiar landscape, with trees of a mixture of ages and origins. On a Saturday in early September, the land was clearly disturbed by humans yet strangely deserted with no farms and virtually no small towns. Cumberland itself, a village of about 400 people, seemed almost a ghost town, hushed by the disappearance of mining jobs. A satellite view still shows the ReCreation Lands as a large beige scar southwest of the village.

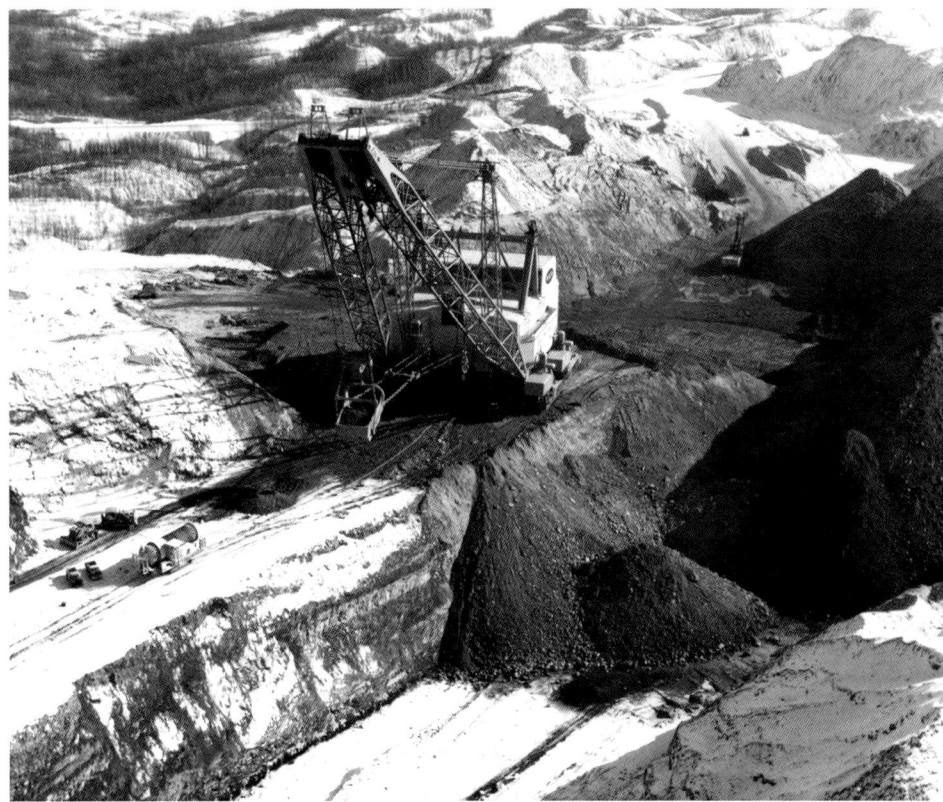

Earthmovers for strip-mining machinery became bigger and bigger after World War II. This photo of the huge walking dragline Big Muskie, once owned by American Electric Power, gives some sense of its tremendous size. The two autos at lower left help to establish scale. Its shovel alone, which today looms over the Miners' Memorial Park in Noble County, has space to accommodate a boy scout troop or a high school marching band. *Tom Root, Ohio Historical Society*

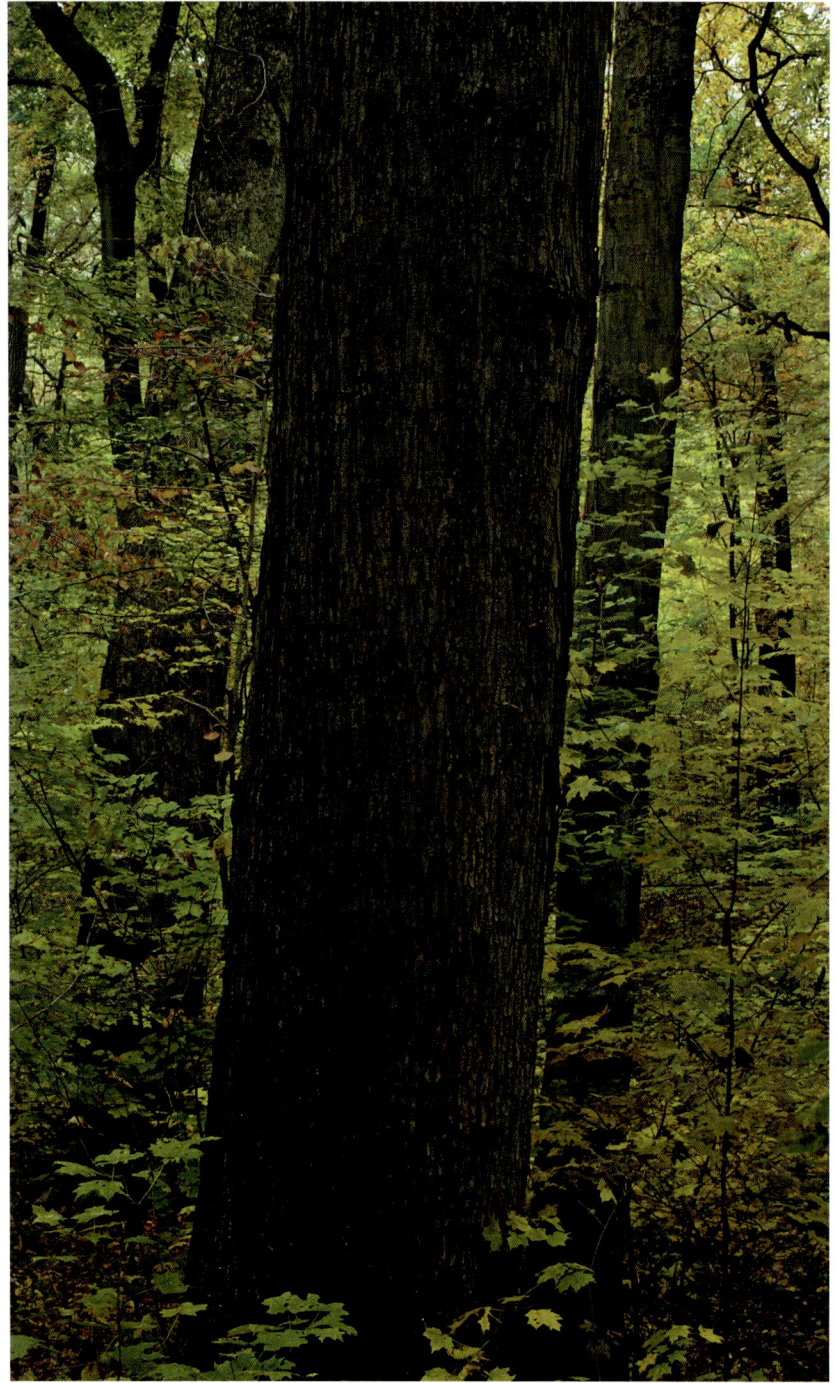

Nearby, at the Wilds, thousands of black locust and autumn olive trees dominate much of the vegetation other than grassland. These trees are often planted to prevent erosion of re-formed land and, in the locusts' case, to replace nitrogen in depleted soil. Unfortunately, they, and other plants, such as crown vetch, are invasives and can suppress regrowth of naturally occuring species; Wilds employees must cut, stack, dry, and burn autumn olive to prevent it from overrunning their acres. Not many miles away, the shovel from Big Muskie sits silhouetted on a ridge at the Miners' Memorial Park above the ReCreation Lands like a stupendous clamshell the size of a bungalow. Mine reclamation is a complex matter. It involves questions such as "Reclamation for what purpose?" and "Reclamation to what degree of historical and ecological authenticity?"

Dysart Woods is perhaps the most significant stand of timber in Ohio. Here, two twenty-acre plots of virgin forest show how Ohio's stately woodlands once looked. Undisturbed except for light grazing and removal of wind-felled trees, the woods were designated a National Natural Landmark in 1967 by the National Park Service. Many trees stand one hundred feet high and measure more than four feet in diameter. *Gary Meszaros*

Worn but Priceless

The hill country's beauty—especially in spring or autumn—sometimes leaves one at a loss for words. Driving through freshly greened pastures along the Flushing Escarpment or gazing down on fiery leaves of a cove forest from the rim of Conkle's Hollow in the Hocking Hills can remind one of paradise. In some places the land may seem eternal, but this is hardly a static Eden from before the biblical Fall. It is an old, worn, and rewoven landscape. Streams laid it down as they ate away lofty mountain ranges raised by the collisions of ancient continents. More streams cut its thick layers into shadowy ravines and sharp ridgetops over further eons. During recent geologic time, retreating continental glaciers thousands of feet thick sent torrents to sculpt it further and filled its valleys with huge finger lakes whose overflow revamped the old circulatory system of rivers and runs.

Plants and animals changed the hill country's face over time, as well, and cove forests, oak-hickory woodlands, and bottomland hardwoods provided contexts for its rich web of life. Most of these processes were, and remain, exceedingly slow, but 200 years ago the pace of change in the hills accelerated like an Indy 500 racer. Within a century, settlers and entrepreneurs had cut nearly all the old-growth forests for farms, iron furnaces,

and railroads and had extirpated many animal species. Coal mines and clay and gravel pits, as well as oil and gas wells, proliferated. Surface mining cut the very tops off many hills. Human needs and

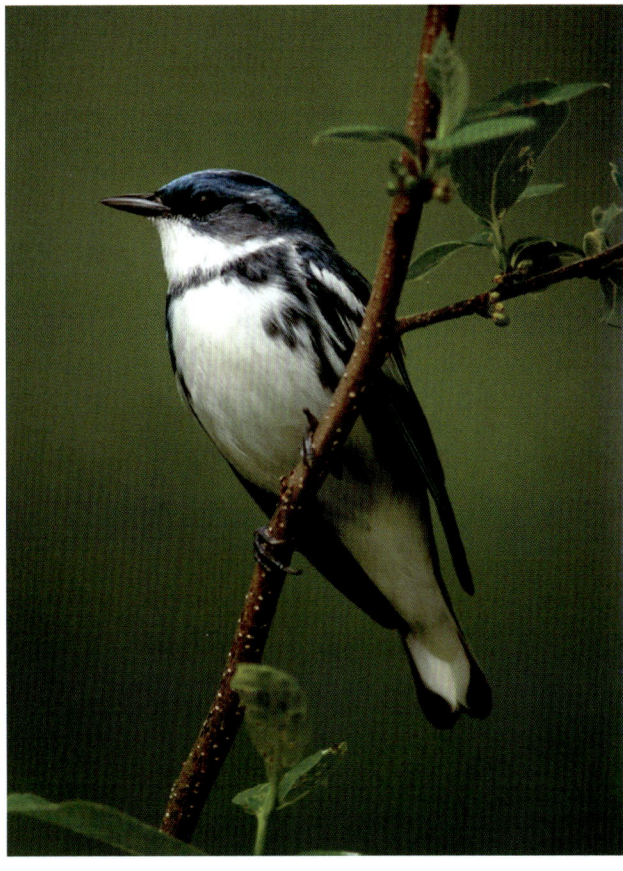

A most elegant warbler, the cerulean is restricted mainly to river corridors where its song drifts down from the canopy's upper level in spring. Loss of mature forest has eliminated this species in many places where it used to be quite common. It is a jewel found along streams which are buffered against agricultural and other runoff, such as high-acid "yellow boy" from mine tailings. *Gary Meszaros*

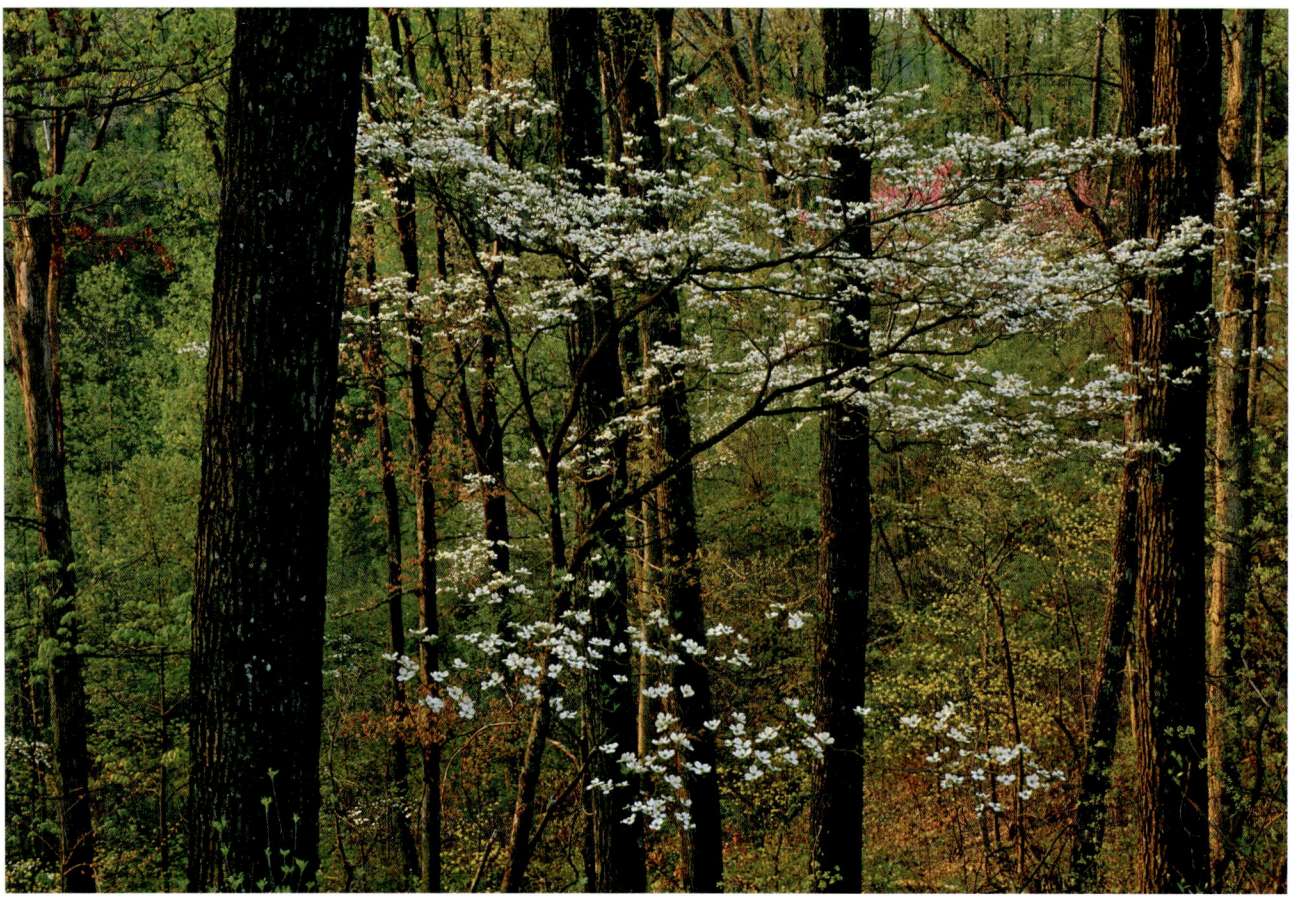

Dogwood is an emblem of spring and probably the showiest flowering shrub in the southern Ohio highlands. *Gary Meszaros*

human-introduced influences—such as the chestnut blight, Dutch elm disease, air and water pollution, development, and climate change—have threatened and continue to threaten area ecosystems. Their effects can be devastating. But the hills, covered again in many places by resurgent greenery, still hold fascination and loveliness for those of us who live on the unglaciated Allegheny Plateau or visit it. This land will continue to enchant us if we remember its history; if we till, mine, and develop it with respect; and if we advocate for its future well-being.

Suggestions for Further Reading

Beatley, Janice B. "The Primeval Forests of a Periglacial Area in the Allegheny Plateau (Vinton and Jackson Counties, Ohio)." *Bulletin of the Ohio Biological Survey, New Series* 1:1. Columbus: Ohio State University, 1959.

Braun, E. Lucy. *Deciduous Forests of Eastern North America.* London: Macmillan, 1950.

———. *The Vegetation of the Mineral Springs Region of Adams County, Ohio. Ohio Biological Survey Bulletin* 15. Columbus: Ohio State University, 1928.

Buchanan, Forest W. "The Breeding Birds of Carroll and Northern Jefferson Counties, Ohio, with Notes on Selected Vascular Plant and Animal Species." *Ohio Biological Survey, Biological Notes No. 12.* Columbus: Ohio State University, 1980.

Camp, Mark J. *Roadside Geology of Ohio.* Missoula, Mont.: Mountain Press Publishing, 2006.

Cusick, Allison W., and Gene M. Silberhorn. "The Vascular Plants of Unglaciated Ohio." *Bulletin of the Ohio Biological Survey, New Series* 5:4. Columbus: Ohio State University, 1977.

Edwards, Clifford D. *Conrad Richter's Ohio Trilogy.* The Hague: Mouton, 1970.

Forsyth, Jane L. "A Geologist Looks at the Natural Vegetation Map of Ohio." *Ohio Journal of Science,* Columbus, May 1970.

Griggs, Robert F. *A Botanical Survey of the Sugar Grove Region. Ohio Biological Survey Bulletin* 3, Columbus, 1914.

Hall, Basil. *Forty Etchings from Sketches Made with the Camera Lucida in North America in the Years 1827 and 1828.* Philadelphia, Pa.: Carry, Lea, and Cary, 1829.

Hansen, Michael C. *Geology of the Hocking Hills State Park Region.* Division of Geological Survey Guidebook No. 4. Columbus: Ohio Department of Natural Resources, 1975.

Howe, Henry L. *Historical Collections of Ohio.* Cincinnati, Ohio: C. J. Krebiel and Co., 1900.

Knepper, George W. *Ohio and Its People.* Kent, Ohio: Kent State University Press, 1989.

Lafferty, Michael B., ed. *Ohio's Natural Heritage.* Columbus: Ohio Academy of Sciences, 1979.

Laycock, George. *The Richard and Lucile Durrell Edge of Appalachian Preserve System, Adams County, Ohio.* Cincinnati, Ohio: Cincinnati Museum Center, 2003.

McCormac, Jim, and Gary Meszaros. *Wild Ohio: The Best of Our Natural Heritage.* Kent, Ohio: Kent State University Press, 2009.

Paleontological Research Institution. *The Teacher Friendly Guide to Geology.* Ithaca, N.Y. http://teacherfriendlyguide.org/.

Platt, Carolyn V. *Creatures of Change: An Album of Ohio Animals.* Kent, Ohio: Kent State University Press, 1998.

Richter, Conrad. *The Trees.* Athens: Ohio University Press, 1991.

Stout, Wilbur, and G. F. Lamb. "Physiographic Features of Southeastern Ohio." *Ohio Journal of Science.* Columbus, March 1938.

Tocqueville, Alexis de. *Democracy in America.* London: Saunders and Otley, 1838.

Williams, Michael. *Americans and Their Forests: A Historical Geography.* Cambridge: Cambridge University Press, 1989.

Wolfe, John N., Richard T. Wareham, and Herbert T. Schofield. "Microclimates and Macroclimate of Neotoma, A Small Valley in Central Ohio." *Bulletin of the Ohio Biological Survey* 8:1. Columbus: Ohio State University, 1949.

Index

forests (cont.)
mesophytic, 19–20, 22; old
growth, *30, 42;* reclaimed after
mining, 41–42; regrowth of,
34–35, 37–38; settlers and, 19,
31–33; temperate, 19; types of,
19. *See also* trees
Fort Hill State Memorial, 28
fossils, 5, 6, 12–13
four-toed salamanders, 25
fragile heel-splitter mollusk, 25
fungi, *36*
fur trade, 37

gas, 3, 6
glaciers: effects of, 1, 16; effects of
melt runoff from, 3, *15,* 16–18,
28; forests before, 19–20; maps
of, *2, 18;* pushing northern
plants into Ohio, 28–29; refor-
estation after, 20
Gondwana, movements of, *11*
grasslands, reclaimed after mining, 41–42
gravel pits, 3, 43
Great Hinckley Hunt, 36–37
green salamanders, 25
Grenville Mountains, 5, *6,* 11
Griggs, Robert, 23
Guernsey County, coal mining in,
41
gypsum, formation of, 8

habitat: destruction of, 35, 37, 43;
of hellbenders, 20; of owls, 21
Hall, Basil, 32–33
Hanging Rock Iron Region, 3,
33–35, *36*
Harrisburg surface, 16
hellbender salamanders, 20

Highland County, Seven Caves
in, 27
hill country: beauty of, 43–44; difference from rest of Ohio, 3–4,
14, 16; threats to ecosystems of,
37–41
Historical Collections of Ohio
(Howe), 19, 34–35, 39
Hocking Hills, 1; cliff rim walk in,
22; forests of, *2,* 21–23; rocks in,
9, *10*
Hocking River, 1
Holmes County, upland forests
in, 23
Howe, Henry, 19, 34–35, 39

Iapetus Ocean, 5
industry, 6, 32–34
inland seas, 6, 9, 11, 16
insects, in swamp forests, 12
Interior Low Plateau, 25, 28
iron industry, 43; consumption of
wood by, 3, 33–35; in Hanging
Rock Iron Region, *36*

Jackson County, forests of, 20, 23
Jefferson County, coal mining in, 39
Jefferson Furnace, 34–35
Jurassic Period, Pangea in, 12–13

Kirtland, Jared, 37

lake plain, in natural regions of
Ohio, *2*
Lake Tight, 17, *18*
lark sparrow, 25
Laurentia, 5–6
Lexington surface, 16
Licking County, forests in, 31
limestone, 16, 26; formation of, 8,

12; in iron production, 34; Silurian, 28
limestone reefs, 8
Little Beaver Creek, *20*
"Little Switzerland," 18
long-tailed salamanders, *24*
lumber/lumbering, 3, 22–23, 33
Lynx Prairie, 29

Mesozoic Age, migration of plants
during, 19
microclimates, 24–25, 28
Miners' Memorial Park, 42
mining: of iron ore, 34. *See also*
coal mining
Mississippian Period, rock formation in, 9–10, *13*
Mohican River, upland forests
along, 23
mollusks, rare, 25
Monroe County, "Little Switzerland" in, 18
moraine, terminal, 1
morels, *36*
mountains, 7; erosion of, *8,* 11; formation of, 11, 13–14; sediments
from erosion of, 6, 9. *See also
specific ranges*
Muskingum Mine, *41*
Muskingum River, 16, 18

National Road, building of, 31
Native Americans, 3, 31–32
natural regions, of Ohio, *2*
Nature Conservancy, 29–30
nature preserves, 25, 27
Nebraskan glacier, 17
Neotoma Valley, microclimates
of, 25
New Straitsville coal mine, 39, *40*

Noble County, giant trees in, 19
North America, movements of, 13
northern white cedar (arbor vitae), *28,* 28–29

Oachita Mountains, formation of,
11
oaks, ancient, 32
Ohio Brush Creek, 25–26
Ohio Department of Natural Resources, forest preserves of, 30
Ohio Historical Society, and Buckeye Furnace, 34, *35*
Ohio River system, 16–18, 28
Ohio Shale, acidic soil on, 26
oil, 6, 43
Ordovician Period: land movements during, *6,* 14; rocks
from, *7, 13*

Pangea, 11–13
Peebles Dolomite, 26
Pennsylvanian Period, 9–10, 12,
13, 16
Permian Period, 11, 16
Perry County, New Straitsville
coal mine in, 39
petroleum industry, 3
Piedmont Terrane, *6*
Pike County, giant trees in, 19
pine trees, 23
plants: animals and, 3; British soldier lichens, 23; Canby's mountain lover, 29; diversity of,
12, 21–22, 25–26, 28; drought-
resistant, 29; on dry ridgetops,
23; ferns, *2;* fire pinks and
wild geraniums, 25; in forest
understories, 22; around Hocking Hills, *10;* influences on, 3,